OXFORD MEDICAL PUBLICATIONS

HUMAN VIROLOGY

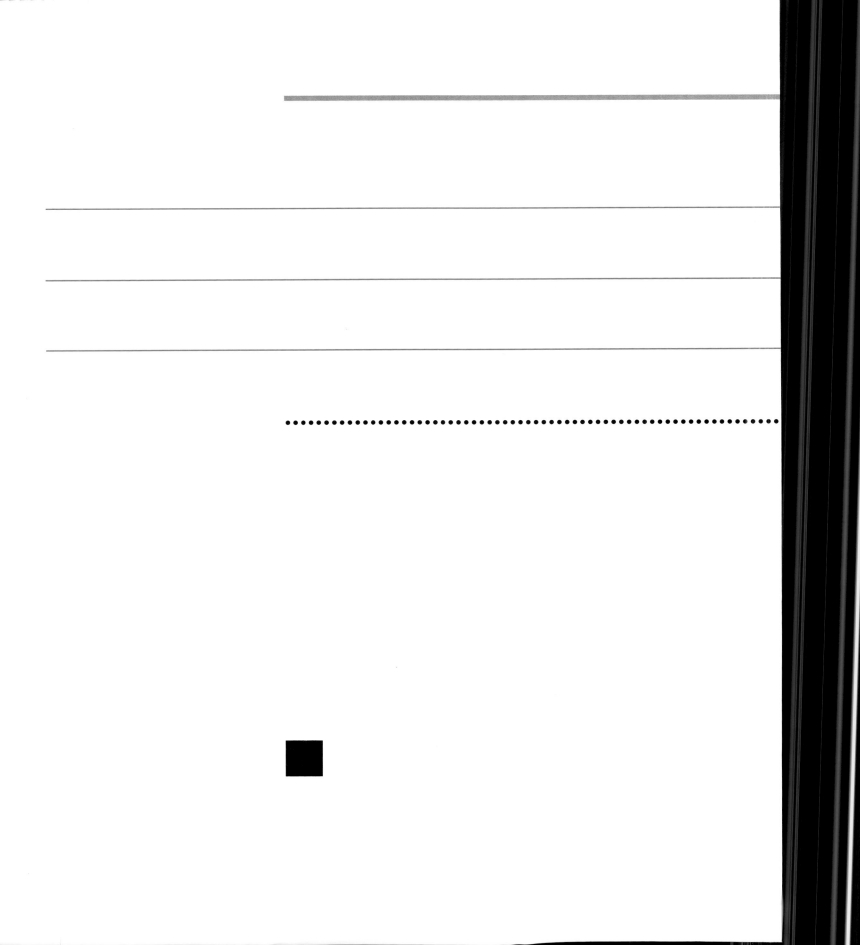

HUMAN VIROLOGY

A TEXT FOR STUDENTS OF MEDICINE, DENTISTRY, AND MICROBIOLOGY

LESLIE COLLIER

Emeritus Professor of Virology
University of London

AND

JOHN OXFORD

Professor of Virology
The London Hospital Medical College

Oxford · New York · Tokyo

OXFORD UNIVERSITY PRESS

1993

Oxford University Press, Walton Street, Oxford OX2 6DP

Oxford New York Toronto
Delhi Bombay Calcutta Madras Karachi
Kuala Lumpur Singapore Hong Kong Tokyo
Nairobi Dar es Salaam Cape Town
Melbourne Auckland Madrid
and associated companies in
Berlin Ibadan

Oxford is a trade mark of Oxford University Press

Published in the United States
by Oxford University Press Inc., New York

British Library Cataloguing in Publication Data
A catalogue record for this book is
available from the British Library

Library of Congress Cataloging in Publication Data
Collier, L. H. (Leslie Harold),
Human Virology. A text for students of
medicine, dentistry, and microbiology
Leslie Collier and John Oxford
(Oxford medical publications)
Includes bibliographical references and index.
1. Medical virology. I. Oxford, J. S. (John Sidney) II. Title.
III. Series.
[DNLM: 1. Virology — methods. 2. Virus Diseases — microbiology.
3. Virus Replication. 4. Viruses — pathogenicity. QW 160 C699v]
QR201.V55C65   1993   616'.0194 — dc20   91-24098
ISBN 0-19-262184-X (cased)
ISBN 0-19-261662-5 (paperback)

Typeset by H Charlesworth & Co Ltd, Huddersfield
Printed in Hong Kong

Preface

In almost every area of biomedical research the making of new discoveries and their subsequent application to the relief of suffering proceed at an ever-increasing pace. Virology, perhaps more than any other discipline, is playing a key role in these advances, not only in the study of infections and their treatment and prevention, but also in the unravelling of the most fundamental aspects of biology. This is because viruses have an intimate relationship with the basic machinery of their host cells. Thus, research on how viruses reproduce themselves and spread has given us many insights into the ways in which the cells of our bodies function, leading in turn to a better understanding of the whole organism and of how infective diseases may be prevented or cured.

For patients, ever-increasing medical knowledge and the improved treatments which it brings are a boon; for students who must absorb more and more information, the benefits may sometimes be less apparent. As teachers of virology, we well appreciate this difficulty. Our aim in writing this book was *not* to try and turn its readers into virologists. It was, first and foremost, to provide information on a 'need to know' basis. But what *do* you need to know of virology? Enough to pass examinations is, for some, the obvious answer. Looking further ahead, in much of medical and dental practice there is a requirement for an intelligent approach to the diagnosis and treatment of viral infections: what are the right questions to ask; what specimens should be taken; what happens to them in the laboratory; what antiviral compounds, sera and vaccines are available? This is the minimum that most doctors and dentists will need in their daily practice and these pages provide it. Some of you, particularly non-clinical students of microbiology, may wish to go into this exciting field in greater depth. We have therefore included more detail about, for example, viral replication and genetics; this will be found in small print in the text and in some of the figure captions, particularly those in Chapters 1 and 2.

The world we live in does not remain static; just as the approach of a new century is bringing fresh political perspectives in Europe and in Africa, reminding us that change and constant questioning are characteristic of human nature, so we should not forget that micro-organisms also inhabit a dynamic and changing world. Our behaviour and alterations in the way we live can dramatically alter their chances for survival and the agents themselves may also mutate and acquire new characteristics. Viruses are constantly being presented with novel opportunities; thus brand-new diseases can emerge with some rapidity. We have witnessed this with AIDS and BSE (bovine spongiform encephalopathy). At the same time, dengue is moving into new territories and influenza A recently revisited the UK 14 years after the last serious epidemic. So, although much of the work of today's virologist is, as before, the repetitious checking of viruses, immune status, and vaccine efficacy, new ideas will always be needed in the fight against a constantly probing and dangerous enemy.

London   L.H.C.
February 1993   J.S.O.

Contents

Abbreviations

Although abbreviations are defined in the text at first mention, a complete list
is given here for ease of reference.

ACV	acyclovir	EBNA	Epstein–Barr (virus) nuclear antigen
ADC	AIDS dementia complex	EBV	Epstein–Barr virus
ADCC	antibody-dependent cellular cytotoxicity	ECHO	enteric cytopathic human orphan (virus)
AIDS	acquired immunodeficiency syndrome	ELISA	enzyme-linked immunosorbent assay
APC	antigen-presenting cell	EM	electron microscope, microscopy
ara A	adenine arabinoside	F	fusion (protein)
ARC	AIDS-related complex	FITC	fluorescein isothiocyanate
ATLL	adult T cell leukaemia/lymphoma	HA	haemagglutinin
AZT	azidothymidine	h	hour(s)
BL	Burkitt's lymphoma	HAM	HTLV-I-associated myelopathy
BMT	bone marrow transplant	HAV	hepatitis A virus
BPV	bovine papilloma virus	HBcAg	hepatitis B core antigen
BSE	bovine spongiform encephalopathy	HBeAg	hepatitis B e antigen
CDC	Centers for Disease Control (USA)	HBIG	hepatitis B immunoglobulin
cDNA	complementary DNA	HBsAg	hepatitis B surface antigen
CFT	complement-fixation test	HCV	hepatitis C virus
CIN	cervical intra-epithelial neoplasia	HDCS	human diploid cell strain (of rabies virus)
CJD	Creutzfeldt–Jakob disease	HEV	hepatitis E virus
CIS	Commonwealth of Independent States (formerly USSR)	HI	haemagglutination inhibition
		HIG	(normal) human immunoglobulin
CMI	cell-mediated immunity	HIV	human immunodeficiency virus
CMV	cytomegalovirus	HLA	human leukocyte antigen
CNS	central nervous system	HPV	human papilloma virus
CPE	cytopathic effect	HRIG	human rabies immunoglobulin
CSF	cerebrospinal fluid	HSV	herpes simplex virus
D and V	diarrhoea and vomiting	HTLV	human T cell leukaemia virus
ddI	dideoxyinosine	IDU	idoxyuridine
DHSS	dengue haemorrhagic shock syndrome	IEM	immune electron microscopy
DI	defective interfering (particles)	IFA	indirect fluorescent antibody (test)
DIC	disseminated intravascular coagulation	IFN	interferon
DNA	deoxyribonucleic acid	Ig, IG	immunoglobulin
ds	double-stranded (nucleic acid)	IL-2	interleukin-2
EA	early antigen (of EBV)	IPV	inactivated polio vaccine

ISCOM	immune-stimulating complex		RT	reverse transcriptase
Kb	Kilobase		S	soluble (mumps antigen)
LCM	lymphocytic choriomeningitis		s	second(s)
LTR	long terminal repeat sequence		SAF	scrapie-associated fibrils
LYDMA	lymphocyte-detected membrane antigen (of EBV)		SIV	simian immunodeficiency virus
			SLA	scrapie-like agents
M	matrix, membrane (protein)		SRH	single radial haemolysis
MAb	monoclonal antibody		SRSV	small round structured virus
ME	myalgic encephalomyelitis		SRV	small round virus
MHC	major histocompatibility complex		ss	single-stranded (nucleic acid)
min	minute(s)		SSPE	subacute sclerosing panencephalitis
MMR	measles–mumps–rubella (vaccine)		STD	sexually transmitted disease
MMTV	mouse mammary tumour virus		SV	simian vacuolating (virus)
MP	mononuclear phagocyte		TAq	*Thermophilus aquaticus* (polymerase)
MRC	Medical Research Council (UK)		Tc	T cytotoxic (cell)
mRNA	messenger RNA		Tdh	T delayed-hypersensitivity (cell)
MS	multiple sclerosis		TFT	trifluorothymidine
NP	nucleoprotein		Th	T helper (cell)
NA	neuraminidase		TK	thymidine kinase
NANB	non-A non-B hepatitis		TME	transmissible mink encephalopathy
NK	natural killer (cell)		TORCH	toxoplasma–rubella–cytomegalovirus–herpes (screening tests)
NPC	nasopharyngeal carcinoma			
OPV	oral (attenuated) polio vaccine		Ts	T suppressor (cell)
PAGE	polyacrylamide gel electrophoresis		TSP	tropical spastic paraparesis
PCP	*Pneumocystis carinii* pneumonia		UV	ultraviolet
PCR	polymerase chain reaction		V	viral (mumps surface antigen)
PHLS	Public Health Laboratory Service (UK)		VAIN	vaginal intra-epithelial neoplasia
PML	progressive multifocal leucoencephalopathy		VCA	viral capsid antigen (of EBV)
RIA	radioimmunoassay		VIN	vulval intra-epithelial neoplasia
RNA	ribonucleic acid		VSV	vesicular stomatitis virus
RNase	ribonuclease		VZV	varicella–zoster virus
RNP	ribonucleoprotein		WHO	World Health Organization
RSV	respiratory syncytial virus		ZIG	zoster immune globulin

PART ONE

GENERAL PRINCIPLES

CHAPTER ONE

GENERAL PROPERTIES AND CLASSIFICATION OF VIRUSES

1 WHAT ARE VIRUSES?

The word virus was used in medical circles as early as 1790, before the germ theory had been formulated; it then simply meant 'poison'. By the end of the nineteenth century, and the climax of the golden age of bacteriology, the causative agents of many infectious diseases had been identified, but pathologists realized that there remained some that were not caused by bacteria or protozoa. Viruses have the following characteristics.

◆ They are **small**, retaining infectivity after passage through filters able to hold back bacteria (Fig. 1.1).

◆ They are **totally dependent upon a living cell**, either eukaryotic or prokaryotic, for replication and existence. Some viruses do possess complex enzymes of their own such as RNA or DNA polymerases but they cannot amplify and reproduce the information in their genomes without assistance.

◆ They possess only **one species of nucleic acid**, either DNA or RNA.

◆ They have a component — a **receptor-binding protein** — for attaching to cells so that they can commandeer them as virus production factories.

Viruses are both subversive and subtle in their operations; in a way, they are the ultimate evolutionary end-point. They have colonized most living beings on this planet, whether plant, animal, insect, or microbe. Their small size and total dependence on a host cell for replication delayed detailed studies of viruses to the present century. Only the techniques of electron microscopy, cell culture, high-speed centrifugation, and electrophoresis allowed their detailed characterization. A major step forward was the discovery of penicillin

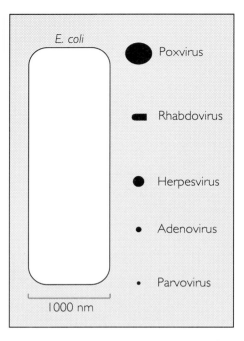

Fig. 1.1 The comparative sizes of representative viruses. The viruses are compared with a typical bacterium (*Escherichia coli*).

and other antibiotics in the 1940s and 1950s; since these compounds inhibit bacteria but not viruses, their addition to cell cultures minimized the risk of contamination and greatly facilitated their use for growing viruses. This advance was brilliantly exploited by the Nobel prizewinners, Enders, Weller, and Robbins, who first grew and studied poliomyelitis and mumps viruses; their researches opened the gate to a flood of discoveries about these and other viruses which is still gathering momentum. During the 1980s the AIDS viruses and new hepatitis and herpes viruses came to light; it is certain that others remain to be discovered.

1.1 *Virus genome structure: positive- and negative-stranded viruses*

Viruses are little more than genetic material packed into a protective shell of protein and there can be no understanding of how they function without a basic knowledge of the chemistry of deoxyribonucleic acid (DNA) and ribonucleic acid (RNA).

Viral nucleic acids carrying genetic information — the **genomes** — can have a number of conformations. The molecules may be double-stranded, as in higher life forms, or single-stranded; they may be linear, circular, continuous, or segmented. Table 1.1 and Figs 1.2, 1.3, and 1.4 show the permutations and give a few examples. We do not want students to memorize all these details, but rather to appreciate how viruses encode the information for manufacture of their proteins.

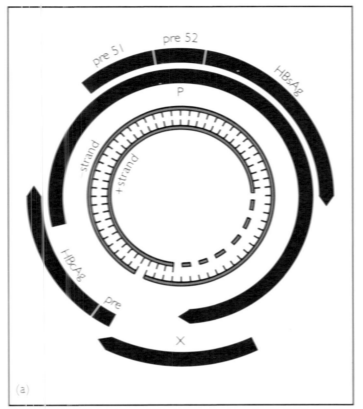

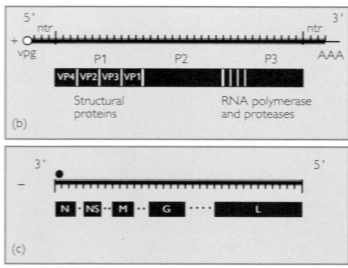

Fig. 1.2 Genome maps of certain DNA and RNA viruses. **(a)** DNA virus (hepatitis B). The circular genome consists of two strands of DNA. The long (negative) strand is complete except for a small nick near the 5′ end of the shorter (positive) strand. The latter is incomplete (represented by dashes) except during replication, when it is completed by an endogenous polymerase (P). Replication is complicated and involves formation by reverse transcription of a hybrid DNA–RNA intermediate.

There are four open reading frames (ORF) on the negative strand: the core antigen (HBcAg) and surface antigen (HBsAg) (see Chapter 20) are encoded in different reading frames, both of which are overlapped by the polymerase (P) ORF. The fourth ORF encodes a protein known as X. **(b)** Positive-strand RNA virus (poliovirus). The virion RNA molecule is polyadenylated (AAA) at the 3′ end and a small virus-coded protein (vpg) is present at the 5′ end. The genome has a single open reading frame whose primary translation product is a polyprotein, which is cleaved to produce viral capsid proteins (VP1, VP2, VP3, and VP4), the RNA polymerase, two proteases, and some minor products. Non-translated regions (ntr) at each end of the genome are not translated into proteins. The 5′ end non-coding region may have a function in the initiation of protein synthesis. **(c)** Negative-strand RNA virus (rabies). A virion-associated polymerase (●) transcribes the 5 genes which are arranged sequentially on the ssRNA genome into 5′ capped, methylated, and polyadenylated mRNAs (not shown) which are translated into the nucleocapsid (N), core phosphoprotein (NS), matrix (M), glycosylated membrane spike (G), and polymerase (L) polypeptides. There is no expansion of information capacity by the use of overlapping reading frames and no spliced mRNAs are found.

It is useful to remember that:

◆ the nucleic acid of all DNA viruses except parvoviruses is double-stranded (but note that hepadnavirus DNA is partly single-stranded when not replicating);

◆ the nucleic acid of all RNA viruses except reoviruses is single-stranded.

In those RNA viruses whose genome consists of single-stranded nucleic acid, the latter is either **positive-stranded** (can act directly as messenger RNA (mRNA)) or **negative-stranded** (must be transcribed by a virus-associated RNA transcriptase enzyme to a mirror-image positive-stranded copy, which is then used as mRNA).

The generation of viral mRNAs, which must compete with cellular mRNAs for the protein-synthesizing machinery of the host cell, is crucial to the replication of all viruses. It is thus not surprising that the concept of positive- and negative-stranded viruses is both central to our understanding of the replication strategy of viruses and important in their classification.

The size of viral genomes varies enormously. The largest, such as those of poxviruses, may contain several hundred genes whereas the smallest may have the equivalent of only three or four.

Oddities among DNA and RNA viruses may occur, in the sense that one often sees by electron microscopy **'empty'** particles which contain no nucleic acid at all. These particles cannot replicate. Others have a defective genome, lacking part of the nucleic acid needed to infect a cell; certain of these are called **defective interfering** (DI) particles, because they interfere with the replication of normal viruses.

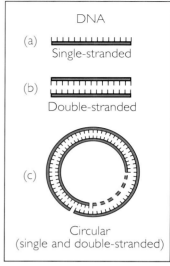

Fig. 1.3 DNA virus genomes.

2 THE STRUCTURE OF VIRUSES

A knowledge of the architecture of viruses is, of course, important in their identification; it also helps us to deduce many potentially important properties of a particular virus. As an example, the processes of attachment and penetration of cells and, later, maturation and release differ greatly according to whether the virus possesses a lipid-containing **outer envelope**. Such enveloped viruses tend to cause **fusion of host cells** at some stage during their entry. A practical consequence of this knowledge might be the design of inhibitors capable of preventing this vital event.

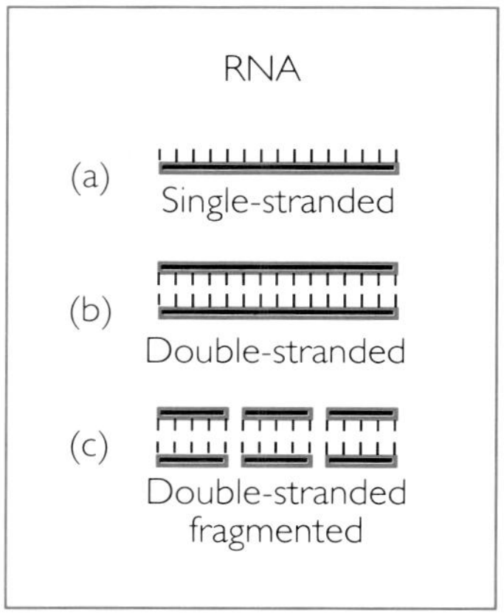

Fig. 1.4 RNA virus genomes.

2.1 *Basic components*

The **protein coat** of a virus is called a **capsid** and is itself made up of numerous **capsomeres**. The function of the capsid is to protect the delicate nucleic acid contained within it. The complex of protective protein and viral nucleic acid is the **nucleocapsid**. Each capsomere may be visualized as a spherical protein, although X-ray crystallography reveals that they are long polypeptide strands woven into complex structural patterns, much like a ball of wool. Some viruses possess an outer envelope containing lipid, derived from the plasma membrane of the host cell during their release by budding from the cell surface. In some such viruses, there is a stabilizing membrane beneath the lipid **bilayer** and also a core structure consisting of protein and the viral genome. The whole virus particle, that is, the nucleocapsid with its outer envelope (if present), is called the **virion**. Most of this knowledge of basic virion structure has come from electron microscope studies.

The vast majority of viruses are divided into two groups, according to whether their nucleocapsids have helical or icosahedral (cubic) symmetry (Table 1.2 and Fig. 1.5). These terms need some explanation.

A spiral staircase is a good example of helical symmetry. If one were to look directly down the centre of a spiral staircase, and if it were possible to rotate it around its central axis, the staircase would continue to have the same appearance; it would be symmetrical about the central axis. In viruses with helical symmetry, the protein molecules of the nucleocapsid are arranged like the steps of the spiral staircase and the nucleic acid fills the central core.

Cubic symmetry is more complicated, but we do not wish to delve too far into three-dimensional geometry! It is enough to say that several solids with regular sides, among them the cube and the icosahedron, share certain features of rotational symmetry, a term referring to the fact that they can be rotated about various axes and still look the same. With few exceptions, all viruses that are not helical are icosahedral, that is, they have 20 equal triangular sides. Because they belong to the same geometric group as the cube, they are often referred to as having cubic symmetry, but we shall use the term 'icosahedral symmetry'. The icosahedral formation is the one that permits the greatest number of capsomeres to be packed in a regular fashion to form the capsid.

Fig. 1.5 Stylized structures of helical and icosahedral viruses. (a) A helical virus. The nucleocapsid is in the form of a spiral staircase, the viral proteins surrounding the nucleic acid. These viruses are often pleomorphic. (b) An icosahedral virus. The 20-sided capsid contains the nucleic acid, which is in a non-helical configuration and may be packaged as a condensed or crystal structure.

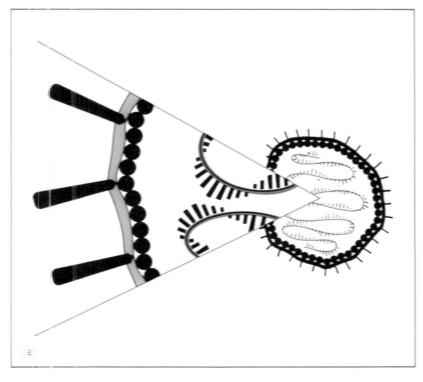

(a)

(b)

2.2 *Viruses with icosahedral symmetry*

These viruses have a highly structured capsid with 20 triangular facets and 12 corners or apices (Fig. 1.6). The individual capsomeres may be made up of several polypeptides, as in the case of poliovirus, in which three different proteins constitute the capsomeres. The capsomeres have a dual function, contributing both to the rigidity of the capsid and to the protection of nucleic acid. Except for the complex poxviruses, all DNA-containing animal viruses have these icosahedron-shaped capsids, as do certain RNA viruses (Fig. 1.7). The DNA-containing herpesviruses have icosahedral symmetry but, in addition, the virion is surrounded by a lipid **envelope**.

2.3 *Viruses with helical symmetry*

Examples of such viruses are the single-stranded RNA viruses such as influenza (Fig. 1.8), the parainfluenza viruses, and rabies (Fig. 1.9). The **flexuous helical nucleocapsid** is always contained inside a **lipoprotein envelope**, itself lined internally with a matrix protein. The lipid of the envelope is derived from the cellular membranes through which the virus matures by budding. Viral glycoprotein spikes project from the lipid bilayer envelope and often extend internally to contact the underlying protein shell referred to as the **membrane**,

Fig. 1.6 (*Below left*) Structural features of adenovirus. Fibres extend from the 12 points, or vertices, of the icosahedral coat, within which is the DNA, illustrated here as a ribbon-like molecule although its precise physical structure is still unknown.

Fig. 1.7 (*Below right*) Structural features of human immunodeficiency virus. The glycoprotein molecules protrude through the lipid membrane. An icosahedral shell underlies the membrane and itself encloses a vase-shaped structure. The diploid RNA is enclosed in the 'vase'.

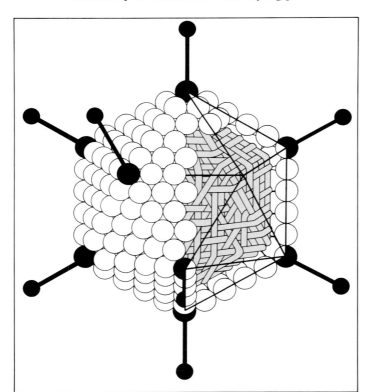

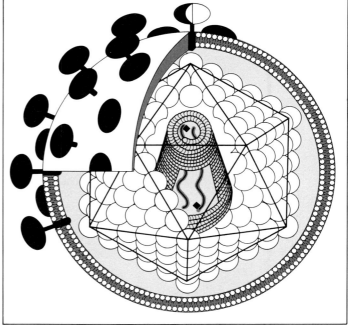

matrix, or **M protein**. This M protein may be rather rigid, as in the case of the bullet-shaped rhabdoviruses, or readily distorted, as in influenza and measles.

2.4 *Complex viruses*

Perhaps not unexpectedly, viruses with large genomes have a correspondingly complex architecture. Poxviruses have lipids in both the envelopes and the outer membranes of the viruses; they are neither icosahedral nor helical, and are referred to by the rather unsatisfactory designation of '**complex**' viruses.

3 CLASSIFICATION OF VIRUSES

The precise pigeon-holing of a virus in a classification system is not only scientifically satisfying, but also of considerable practical consequence. A single example illustrates this assertion. The AIDS virus (HIV-1) was at first thought to belong to the tumour virus group of the family *Retroviridae* (Chapter 6). When, by examination of its detailed morphology and establishment of its genome structure, HIV-1 was shown to be more related to the lentivirus group (Chapter 22), many previously unknown features of its biology could immediately be filled in by reference to other known viruses of the same group. Viral classification need not be a dry, academic subject.

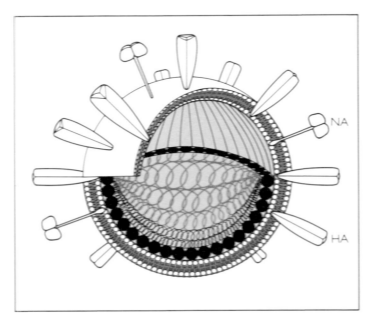

Fig. 1.8 Structural features of influenza virus. The spikes of haemagglutinin (HA) and neuraminidase (NA) protrude from the lipid bilayer; beneath this is a layer of M protein, which in turn encloses the segmented RNA genome, each segment of which is covered with the nucleocapsid protein.

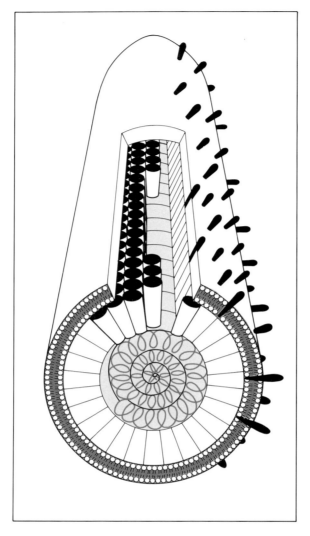

Fig. 1.9 Structural features of rabies virus. Glycoprotein spikes protrude through the lipid of the bullet-shaped virion. The M protein encloses the RNA, which is closely associated with the nucleoprotein. The N protein subunits are represented as short cylinders.

The following are the main criteria used for the classification of viruses:

◆   the **type of nucleic acid** (DNA or RNA);

◆   the **number of strands of nucleic acid** and their physical construction (single- or double-stranded, linear, circular, circular with breaks, segmented);

◆   **polarity of the viral genome**. RNA viruses in which the viral genome can be used directly as messenger RNA are by convention termed 'positive-stranded' and those for which a transcript has first to be made as 'negative-stranded';

◆   the **symmetry** of the nucleocapsid;

◆   the presence or absence of a **lipid envelope**.

On these criteria, viruses are now grouped into **families**, **subfamilies**, and **genera**. Further subdivision is based on the degree of antigenic similarity. Classification lacks precision beyond this point, but in certain circumstances antigenically identical viruses can be further categorized by differences in biological characteristics such as virulence, or in chemical structure, for example, their nucleotide sequences. Tables 1.1 and 1.2 summarize the biological and structural characteristics of the families of viruses known to cause disease in humans.

A cornucopia of information about nucleotide sequences, the organization of viral genomes, and the structures of the protein molecules they encode is available and has already influenced our older and more established ideas about classification of viruses. Modern analytical methods have also unearthed some interesting relationships, including amino-acid sequences common to retroviruses, hepadnaviruses, and cauliflower mosaic virus, and the related thymidine kinases of human and chicken cells and three orthopoxviruses. These resemblances presumably indicate common ancestries in the remote past or common functions. Of course, in the absence of fossil remains of viruses, calculations of rates of change of

Table 1.1 Characteristics of viral nucleic acids and properties related to classification

Family name	Genome*	Positive or negative stranded genome†	Infectivity of isolated nucleic acid	Polymerase or transcriptase present in virion
DNA viruses				
Parvoviridae	ss	±	+	None
Papovaviridae	ds	±	+	None
Adenoviridae	ds	±	+	None
Herpesviridae	ds	±	+	None
Poxviridae	ds	±	−	DNA–RNA
Hepadnaviridae	ds	±	?	RNA–DNA
RNA viruses				
Picornaviridae	ss	+	+	None
Flaviviridae	ss	+	+	None
Togaviridae	ss	+	+	None
Coronaviridae	ss	+	+	None
Bunyaviridae	ss	−	−	RNA–RNA
Orthomyxoviridae	ss	−	−	RNA–RNA
Paramyxoviridae	ss	−	−	RNA–RNA
Rhabdoviridae	ss	−	−	RNA–RNA
Arenaviridae	ss	−	−	RNA–RNA
Retroviridae	ss	+	−	RNA–DNA
Reoviridae	ds	−	−	RNA–RNA
Filoviridae	ss	−	−	RNA–RNA
Caliciviridae	ss	+	?	None

*ss, Single-stranded; ds, double-stranded
† ±, Strands of both polarities present.

Table 1.2 Important structural characteristics of the families of viruses of medical importance

Family name	Representative viruses	Approximate diameter of virion (nm)	Symmetry of nucleocapsid
DNA viruses			
Parvoviridae	Human parvovirus B19	20	I
Papovaviridae	Wart viruses	50	I
Adenoviridae	Adenoviruses	80	I
Herpesviridae	Herpes simplex viruses	180	I
Poxviridae	Vaccinia virus	250	C
Hepadnaviridae	Hepatitis B virus	40	I
RNA viruses			
Picornaviridae	Polioviruses	25	I
Flaviviridae	Yellow fever virus	30	I
Togaviridae	Rubella virus	80	I
Coronaviridae	Infectious bronchitis virus	100	H
Bunyaviridae	California encephalitis virus	100	H
Orthomyxoviridae	Influenza viruses	100	H
Paramyxoviridae	Measles virus	150	H
Rhabdoviridae	Rabies virus	150	H
Arenaviridae	Lassa fever virus	100	H
Retroviridae	HIV–I	100	I
Reoviridae	Rotaviruses	70	I
Filoviridae	Marburg virus	Variable	H
Caliciviridae	Calicivirus	35	I

I, icosahedral; H, helical; C, complex.

viral genes must remain as estimates. Viruses evolve rapidly because (1) they undergo many genome duplications in a short time (adenovirus, for example, may produce 250 000 DNA molecules in an infected cell); (2) the host cell has evolved an editing system to test for mismatched base pairs and viruses may be unable to perform this function. With RNA viruses the rate of divergence of RNA genomes at the nucleotide level can be as high as 2.0 per cent per year. This is a million times the rate for eukaryotic DNA genomes. Viral RNA polymerases have a very high frequency of error.

4 NOMENCLATURE OF VIRUSES

The nomenclature of viruses has been in a state of flux for many years, and the names now in general use are based on characteristics that vary from family to family. Some viruses are named according to the type of disease they cause, e.g. poxviruses and herpesviruses; other familial names are based on acronyms, e.g. papovaviruses (*papilloma–polyoma–vacuolating* agent) and picornaviruses (*pico*: small; *rna*: ribonucleic acid); others again are based on morphological features of the virion, e.g. coronaviruses, which have a halo or corona of spikes. Some individual viruses are named after the place where they were first isolated, e.g. Coxsackie, Marburg, or occasionally after their discoverers, e.g. Epstein–Barr virus.

The official nomenclature of families and subfamilies is latinized, e.g. *Picornaviridae*, *Poxviridae*. A few families are divided into subfamilies, with the suffix *-virinae*, e.g. *Chordopoxvirinae*. Genera are not latinized, e.g. Orthopoxvirus, Parapoxvirus. This system is rather heavy going, so we shall use the appropriate colloquial names (picornaviruses, poxviruses), except where use of the latinized versions is necessary on taxonomic grounds.

5 THE RANGE OF DISEASES CAUSED BY VIRUSES

We have so far dealt with viruses only as micro-organisms; we shall now start to look at them in relation to the diseases that they cause in humans and, where relevant, in animals.

Viruses vary greatly not only in their **host range**, but also in their affinity for various tissues within a given host (**tissue tropism**) and the mechanisms by which they cause disease (**pathogenesis**). Some viruses are

Table 1.3 Relationship between some families of viruses and the range of syndromes that they cause

Syndrome	Virus family			
	Picornaviridae	Togaviridae	Rhabdoviridae	Hepadnaviridae
Febrile illness only	+	+		
Skin rashes	+	+		
Conjunctivitis	+			
Respiratory illness	+			
Diarrhoea	+			
Myositis	+	+		
Arthritis		+		
Meningitis	+	+		
Paralysis	+			
Hepatitis	+	+		+*
Rubella		+		
Rabies			+	

* Can also cause arthritis and rash in the prodromal stage.

predominantly neurotropic, others replicate only in the liver or skin, and others again can infect many of the body systems. These variables are reflected in the ranges of diseases caused by viruses of different families. Despite their similarity in properties, the members of some families cause a wide variety of **syndromes** (sets of symptoms and physical signs occurring together); by contrast, the range of syndromes caused by other families is much more restricted. Table 1.3 gives some examples. The picornaviruses cause many types of illness, the togaviruses somewhat fewer, and the hepadnavirus and rhabdovirus families cause disease only in the liver and brain, respectively.

Table 1.3 also shows that there may be considerable overlap between the symptoms caused by very different viruses; as may be imagined, this can create problems in diagnosis, most of which can, however, be solved by studying the clinical picture as a whole and enlisting the aid of the virology laboratory as early in the illness as possible.

6　REMINDERS

◆　Viruses are characterized by their:
1. small size;
2. obligate intracellular parasitism;
3. possession of either a DNA or an RNA genome;
4. possession of a receptor-binding protein.

◆　The genome is protected by a coat or capsid consisting of protein subunits (**capsomeres**). Each capsomere is made of 1–3 polypeptides. Some viruses have a **lipid envelope**. The nucleic acid is often complexed with a protein. The nucleic acid core and the capsid are together known as the **nucleocapsid**. The complete virus particle (nucleocapsid with **envelope**, if present) is termed the **virion**.

◆　The nucleocapsids of nearly all viruses are built either like **helices**, with the capsomeres arranged like the steps in a spiral staircase around the central genome core, or like **icosahedra**, in which the capsomeres are arranged to form a solid with 20 equal triangular sides, again enclosing the genome. These forms are referred to, respectively, as having helical or icosahedral (cubic) **symmetry**. The large poxviruses have complex structures not falling into either category.

◆　Viruses are classified into **families** according to the characteristics of their nucleic acid, whether DNA or RNA, the number of strands, and their polarity. Positive-stranded viruses use the viral genome as mRNA. Negative-stranded viruses must make a transcript of the genome RNA to be used as a message.

◆ Some families are divided into **subfamilies** on the basis of gene structure. Further subdivisions into **genera** depend on antigenic and other biological properties.

◆ Viruses vary widely in the range of hosts and tissues that they can infect; members of some families cause a wide range of syndromes, whereas the illnesses due to others are much more limited in number.

CHAPTER TWO

VIRAL REPLICATION AND GENETICS

1 INTRODUCTION

In the task of reproducing themselves, viruses are at a major disadvantage compared with higher forms of life. The latter all multiply by some form of fission, so that the daughter cells start their existence with a full complement of genetic information and with the enzymes necessary to replicate it and to catalyse the synthesis of new proteins. A virus, on the other hand, enters the cell with nothing but its own puny molecule of nucleic acid, and sometimes without even a single enzyme to start the process of replication. This is why it must rely so heavily on the host cell for the materials it needs for reproduction and why the replication of viruses is more complicated in some respects than that of other micro-organisms. Although we now have a fairly detailed picture of the main steps, we still do not know everything about the strategies that viruses have developed over the millenia to continue their existence.

In the following description, we shall often use two key terms.

Transcription is the copying of a DNA molecule to make a complementary RNA molecule; the process is somewhat analogous to making a photographic print from a negative. RNA may be transcribed into messenger RNA (mRNA) with a transcriptase enzyme or into DNA with a reverse transcriptase.

Translation, on the other hand, is the reading of the triplet genetic code of nucleotide bases on the mRNA which in turn leads to the linking in the correct sequence of amino acids to form polypeptides (proteins).

2 EVENTS LEADING TO INFECTION OF THE HOST CELL

After the initial infection of a cell, there follows a period of a few hours during which nothing seems to be happening. This appearance is, however, deceptive because much is going on at the molecular level, such as transcription of the 'incoming' viral genes. Figure 2.1 illustrates a fundamental difference between the replication of viruses and bacteria; the latter retain their structure and infectivity throughout the growth cycle, whereas viruses lose their physical identity and most or all of their infectivity during the initial stage of replication, which for this reason has been termed the **eclipse phase**. The next stage, the **productive phase**, is even more full of action as new virus particles are produced and released from the cell.

2.1 *General plan of viral replication*

You are probably familiar with the mechanisms whereby DNA is replicated and transcribed to messenger RNA (mRNA), which in turn is translated to new proteins. In many viruses, however, the genetic information is encoded in RNA and their modes of replication are rather different. We shall describe them a little later, but first, let us look at the main steps in the replication of a DNA virus (steps numbered as in Fig. 2.2). The clocks give an approximate indication of the time-scale.

(1) The virus adsorbs to a host cell. (2) A few minutes later it has entered the cell after which (3) it 'uncoats', i.e. sheds its protective protein

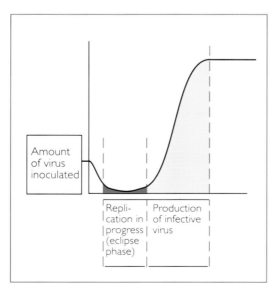

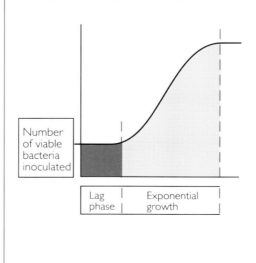

Fig. 2.1 Typical growth curve of a virus compared with that of a bacterium.

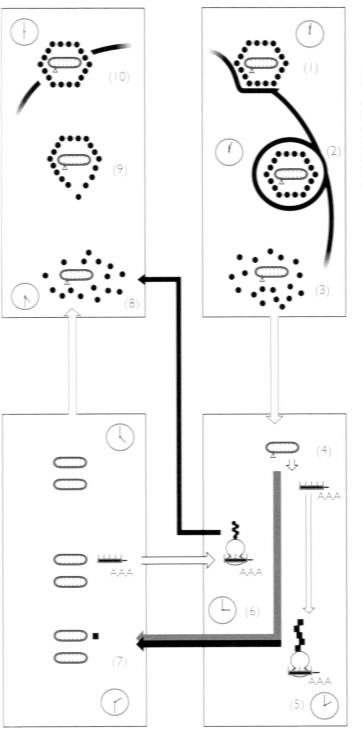

Fig. 2.2 Life cycle of a DNA virus (vaccinia). This, unlike other DNA viruses, replicates exclusively in the cytoplasm. Initial transcription takes place in the core of the virion; the protein products of the transcripts function to release the viral genome from the core. The genomic DNA strands are covalently linked at their ends. For further explanation, see text.

△	Virus associated DNA-dependent RNA polymerase
●	Late, mainly structural proteins
■	Early proteins including DNA-dependent DNA polymerase

shell. (4) The DNA is transcribed to various mRNAs, which code for (5) 'early' or (6) 'late' proteins, both of which are produced on the ribosomes of the host cell. The late mRNAs are transcribed from newly synthesized DNA. The early proteins may have various functions; for example, some are polymerases that catalyse the synthesis of new DNA molecules (7) and others are transcriptional activators. The 'late' proteins are mostly structural polypeptides (8) that are (9) assembled with the new DNA to form progeny virions. These are then released from the cell (10) the whole process having taken 6–8 h. The new infective virions are then free to infect neighbouring cells and start the process over again.

We shall now describe these steps in more detail, and explore the variations on the theme adopted by RNA viruses.

2.2   *Recognition of a 'target' host cell*

All viruses have on their outsides a protein containing a **receptor-binding site** which often takes the form of a pocket or a protuberance that reacts specifically with a corresponding receptor on a cell surface. This precise key-and-lock interaction explains why many viruses are restricted to a given host and, within that host, to particular tissues. Thus the AIDS virus, HIV-1, recognizes and reacts specifically with a receptor on certain T lymphocytes and other cells and can thus attach to and infect only these cells. Likewise, poliomyelitis virus can bind only to the cells of primates and will not infect those of other species. The receptors on cells are glycoproteins or glycolipids.

2.3   *Internalization of the virus*

Having attached to the host cell, the virus must penetrate the external plasma membrane and release its genome into the cellular milieu for subsequent replication. This is accomplished in one of three ways:

◆   **fusion** of the viral membrane at the external cellular plasma membrane and subsequent release of viral nucleic acid;

◆   internalization of the whole virion by **viropexis** (or pinocytosis) and subsequent fusion with an internal vacuolar membrane to release viral nucleic acid;

◆   viruses without a lipid membrane appear to pass or slide through the external plasma membrane directly.

Fusion at the cellular external plasma membrane ('fusion from without') is the strategy of entry of paramyxoviruses such as measles and mumps viruses (Fig. 2.3(a)). Such viruses have a **'fusion protein'** with a short stretch of hydrophobic amino acids which mediates fusion between the lipids of the virus and the cell membrane. The fusion sequence has to be in contact with both viral and host-cell lipid membranes to initiate the joining event.

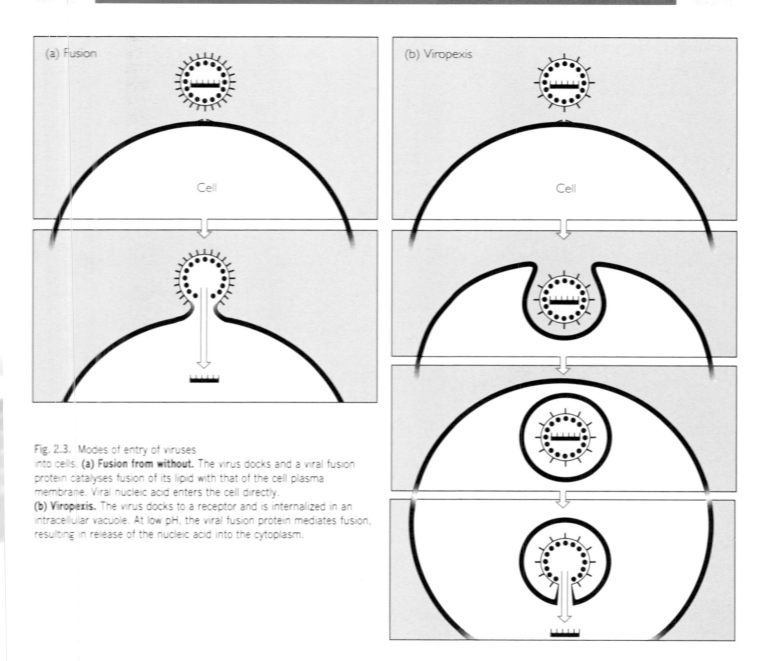

Fig. 2.3. Modes of entry of viruses
into cells. **(a) Fusion from without.** The virus docks and a viral fusion
protein catalyses fusion of its lipid with that of the cell plasma
membrane. Viral nucleic acid enters the cell directly.
(b) Viropexis. The virus docks to a receptor and is internalized in an
intracellular vacuole. At low pH, the viral fusion protein mediates fusion,
resulting in release of the nucleic acid into the cytoplasm.

Viropexis is a more common cellular entry technique and is a 'back door' approach
(Fig. 2.3(b)). Mammalian cells have had to develop methods of attachment
and entry of a range of essential molecules such as nutrients and hormones.
Viruses can exploit this existing avenue of entry and attach at special areas
on the cell membrane where the protein **clathrin** is present. The clathrin
forms a so-called **coated pit** and, once the virus has attached, inversion of
the cellular membrane and associated virus occurs. The virus is now in the

cytoplasm but is still bounded by the cell membrane through which it has to negotiate a route to the true internal environment of the cell. Some viruses, such as influenza, achieve this by an **internal fusion** ('fusion from within') mediated by the viral haemagglutinin (HA) protein. A further requirement of internal fusion is a low pH in the cytoplasmic vacuole, which triggers a movement of the three-dimensional structure of the HA protein, so allowing juxtaposition of the fusion sequence with both lipid membranes. Once 'internal' fusion has been accomplished, the viral nucleic acid is freed into the cytoplasm and, in the case of most viruses, migrates rapidly to the cell nucleus for subsequent transcription and replication. It is a mystery how the viral nucleic acid, particularly single-stranded RNA, protects itself from destruction by the many nuclease enzymes present in the cytoplasmic vacuole.

2.4 *Transcription of viral genome to mRNA and replication of the viral genome*

Positive- and negative-stranded RNA viruses

RNA viruses have three transcription strategies, which depend upon the message sense of their genome RNA. In the so-called **'positive-strand' RNA viruses**, the viral RNA is the messenger RNA and can be translated into viral proteins by cell ribosomes directly the virus enters the cell (Fig. 2.4).

In this paragraph, the numbers in parentheses refer to those in Fig. 2.4.

The positive-stranded parental viral RNA (1), with the addition of a poly-A (AAA) tract at one end of the molecule, is used directly as viral mRNA, from which 'early' and 'late' viral proteins (see Section 2.1) are translated directly (2). Replication of the parental genome occurs by synthesis of a double-stranded intermediate form (3) from which many copies (4) of the original parental genome are made; this process is mediated by the newly translated RNA polymerase.

In the case of the **negative-stranded RNA** viruses, e.g. influenza, a virus-associated RNA polymerase must first make mirror-image copies of the original viral RNA segments (1) and these copies then act as mRNAs (2).

Fig. 2.4. Replication strategy of a positive-stranded RNA virus (poliovirus). The genomic RNA acts directly as mRNA and is translated to give a polyprotein which is rapidly cleaved by two virus-coded proteases into 12 or more smaller proteins (not all illustrated). At a later stage during replication the number of positive RNA strands increases and these are used either as mRNAs or are packaged into virions. By contrast with poliovirus, other positive-strand RNA viruses, e.g. coronaviruses, utilize mRNA molecules of subgenomic length. Coronaviruses also use translational frameshifting to extend the information encoded on certain of the mRNAs.

Fig. 2.5. (*Below right*) Replication strategy of a negative-stranded RNA virus (influenza). The viral genome is in the form of eight loosely linked single-stranded RNA segments. Most transcribed mRNAs are monocistronic, i.e. they code for a single protein. However, the mRNAs of genes 7 and 8 have undergone splicing and each now codes for 2 viral proteins. The mode of transcription of influenza virus is unique since both host and virus RNA polymerases are utilized.

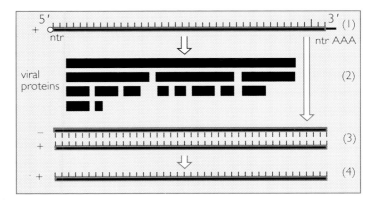

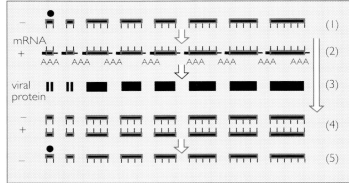

(Numbers refer to steps in Fig. 2.5.) mRNAs are then translated to give viral proteins (3). RNA polymerases transcribe the parental negative-stranded RNA (1) into a complete copy in the form of a double-stranded intermediate (4). New negative-stranded RNAs (5) for incorporation into virions are synthesized from the intermediate form.

The third group of RNA viruses — the **retroviruses** — have an even more complex replication strategy. The essentials of replication and integration are illustrated in Chapter 6, Fig. 6.2.) The parental viral RNA is transcribed by a virus-associated reverse transcriptase (RNA-dependent DNA polymerase) into an ssDNA copy, which forms a DNA–RNA hybrid. The RNA strand is digested away and replaced by a DNA copy to give a dsDNA molecule. This is integrated into the chromosomal DNA of the host cell and is now termed proviral DNA. Viral mRNAs are transcribed from the proviral DNA and viral proteins are synthesized. Complete copies of viral RNA to be incorporated into new virions are also transcribed from the proviral DNA.

DNA viruses

Figure 2.7 gives the replication strategy of a **DNA virus**; numbers in this paragraph refer to this figure. Viruses containing DNA (1) must also produce an mRNA transcript (2) soon after the infection of a cell. This is usually achieved by a host-cell enzyme, DNA-dependent RNA polymerase II. Again, the viral mRNAs are translated to give 'early' and 'late' proteins (3). Early mRNAs are transcribed from input parental DNA, whereas late mRNAs are transcribed from newly replicated DNA (see Fig. 2.2). The parental dsDNA is usually replicated by a virus-coded DNA-dependent DNA polymerase. The mechanisms include priming by a protein linked to the 5′ end of each strand

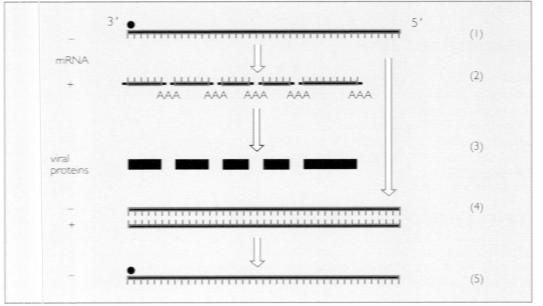

Fig. 2.6. Replication strategy of a negative-stranded RNA virus (rabies). The viral genome is in the form of a single complete strand of RNA. Five mRNAs are transcribed by a start-and-stop mechanism and each is translated into a viral protein.

● RNA-dependent RNA polymerase

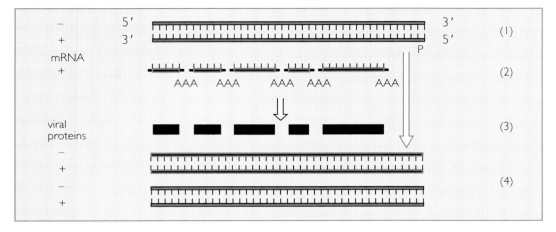

Fig. 2.7. Replication strategy of a DNA virus (adenovirus). The genomes of adenovirus, papovavirus, papillomavirus, and herpes viruses are transcribed and replicated in the cell nucleus. The replication of the linear adenovirus DNA duplex molecule is mediated by a protein (P) at the 5′ end of each DNA strand. The viral genome retains its linear configuration throughout infection. Multiple mRNAs (not all shown) are transcribed from both DNA strands. The genome is large and codes for many proteins. Early mRNAs are encoded by input parental DNA. Late mRNAs are encoded predominantly in the same DNA strand and have a common non-coding leader sequence at the 5′ end containing signals for initiation of transcription and messenger capping. Splicing is extensively utilized and can provide control of expression of different regions of the genome, as well as a means of changing the reading frame.

of a DNA molecule (adenovirus), a rolling circle mechanism (vaccinia), and utilization of a replication fork (papilloma) with synthesis of short fragments of DNA on one strand which are then ligated to form a complete molecule, and continuous copying on the second strand.

2.5   *Processing of messenger RNA*

Primary mRNA transcripts may be processed before they are translated into proteins. mRNA splicing is particularly common among DNA viruses (e.g. adenovirus) but some retrovirus and influenza virus mRNAs are also spliced. A second essential feature of mRNA processing is the addition of a sequence of 50–200 adenylate residues to the 3′ terminus of the molecule. A 'cap' of 7-methyl guanosine is added at the 5′ terminus and helps to form a stable initiation complex on the ribosome.

2.6   *Synthesis of viral proteins*

Viral mRNAs are translated on normal host-cell polyribosomes to produce viral structural and non-structural proteins, which are synthesized very rapidly, in a matter of a few minutes. Host-cell enzymes are responsible for reading the genetic message of the virus, in a triplet code with start and stop

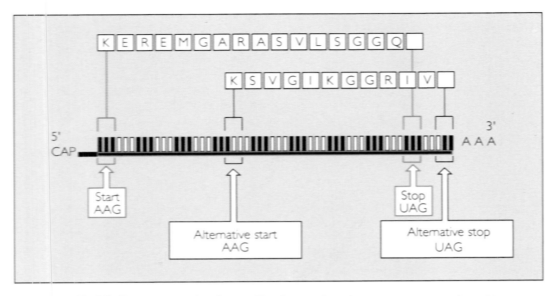

Fig. 2.8. Overlapping reading frames. The diagram shows how two—or more—polypeptides can be synthesized from a single length of nucleic acid. Starting at an initiation codon AUG, the nucleotide sequence is read to give a protein MLRFMG....., etc. An alternative start codon situated farther down the nucleic acid molecule allows a different protein, MVTAYS....., etc. to be translated from the same gene, since the nucleotide triplets are now read in a different sequence. The boxed letters are the conventional codes for the various amino acids: M, methionine; V, valine; L, leucine, etc. For clarity, the base triplets are represented alternatively as open and solid bars; the shading has no other significance.

codons, as in a normal uninfected cell. The structural proteins are the building proteins of the virus particle, whereas the non-structural proteins have important functions, often enzymatic, but do not usually become part of the virion itself. With certain viruses, e.g. herpes, viral protein synthesis is precisely controlled at the level of transcription in quantity and in time.

More often than not, a single viral protein is synthesized from a single mRNA. Some viruses have a second strategy, whereby a very large viral polyprotein is translated initially from a single viral messenger mRNA. This polyprotein is then cleaved at specific sites by viral or cellular proteolytic enzymes to give a series of smaller viral proteins, some of which will be incorporated into the virus. In a third strategy, two virus proteins may be encoded by a single mRNA since the mRNA may be read in different reading frames (see Fig. 2.8). Retroviruses such as HIV-1 use **ribosomal frameshifting**, whereby the ribosome 'jumps' at a particular point and thus begins reading the triplet code in a different frame.

Once synthesized, the individual virus structural proteins migrate to the plasma membrane (budding viruses) or, alternatively, may assemble in the cell cytoplasm (lytic viruses) or the nucleus. In either case, carbohydrate is added to some of the newly synthesized viral proteins by cellular enzymes and any further processing such as addition of acyl or other chemical side chains is also accomplished by cellular enzymes.

2.7   *Release of virus from the host cell*

The virus is now nearing the time of maturation and release. Its structural proteins have been synthesized by a host cell that appears relatively normal, or that may be irretrievably damaged. These viral proteins have been transported by the existing cell machinery and have, in the case of 'budding' viruses, been inserted in the **external plasma membrane** of the cell. Other virus structural proteins also migrate to the inside of the plasma membrane. The proteins and nucleic acid **assemble** and bud by protrusion through the cellular plasma membrane. The bud is pinched off and a new virus is born. Often with budding viruses, such as influenza and measles, the cell may continue to produce successive waves of new viruses. As many as 10 000 virions may be produced per cell in as few as 6 hours, so that we can readily appreciate the potential for rapid spread and colonization. Even at the prebudding stage a virus becomes vulnerable to **immune surveillance** (Chapter 5). The virus is even more vulnerable to host defences and attack once it has budded. Some of these viruses do not emerge from the cell, but may spread to contiguous cells via **connecting pores** or by fusing their membranes.

Some viruses, such as poliovirus, may assemble completely in the cytoplasm and are released only after lysis and death of the cell.

3   GENETIC VARIATION OF VIRUSES

Mutation events such as removal or insertion of a nucleotide or a group of nucleotides (deletion or insertion mutants) are not uncommon during virus replication and are much more frequent in RNA than in DNA viruses. In fact, all RNA viruses are thought to exist as mixtures of countless genetic variants with slightly different genetic and antigenic compositions. These mixtures are dynamic within the host, so that under a particular set of conditions one virus is dominant but others are still present, albeit in much lower numbers. Subtle pressures, such as development of specific immunity to a particular variant, provide the power for viral evolution.

Another way in which viruses may vary their genomic structure is by **recombination**, which is brought about by the exchange and subsequent covalent linkage of genome fragments from a single gene or from two infecting viruses of the same kind. Fortunately, such genetic interactions do not occur among unrelated viruses (such as polio and influenza), otherwise our problems would be greatly compounded! This type of genetic interaction may give rise to a virus with hitherto unknown characteristics and may also give it a selective advantage over its relatives. More often though, the recombinant will have properties incompatible with survival.

With certain RNA viruses, such as influenza and rotaviruses, in which the genome exists as separate fragments, simple **exchange** of genes may occur, a process known as gene **reassortment**. Such reassortant viruses have characteristics that differ from those of the parental viruses. The frequency of such gene exchanges may be very high, much higher, for example, than that of true recombination. Such genetic reassortment can extend the gene pool of the virus and allow the emergence of new and successful variants.

4 REMINDERS

◆ The stages of viral infection of cells are **cellular recognition, internalization, genome transcription** to form mRNA, **mRNA translation, genome replication**, and release from the cell. The complete cycle characteristically takes 6–8 h.

◆ Some RNA-containing viruses such as polio are **positive-stranded** and **their genome acts directly as mRNA**. **Negative-stranded RNA viruses**, such as influenza, possess a **virion-associated RNA transcriptase** that produces an **mRNA transcript** from the genome RNA.

◆ Transcription of **DNA viruses** is carried out by cellular **DNA-dependant RNA polymerases**.

◆ Viral mRNAs may be **spliced**, thus allowing several messages to be carried in a single piece of genome. They may also be **read in different reading frames** at the translation stage, again allowing more extensive use of genetic information.

◆ Viruses may **bud** from infected cells in many waves or may be released instantaneously by **cell lysis**.

◆ The replication of RNA viral genomes is error-prone; this generates genomic diversity. Replication of DNA viruses is semi-conservative and extremely faithful, resulting in fewer genetic changes.

◆ **Genetic recombination** and gene reassortment may both lead to **genetic diversity**.

CHAPTER THREE

PROPAGATION OF VIRUSES IN THE LABORATORY

1 INTRODUCTION

As with other microbes, we need to grow viruses in the laboratory for three main reasons: **diagnosis** of infection; **research**, including infectivity titrations; and, on a larger scale, **production of antigens** for vaccines and serological reagents.

In this context, two important characteristics of viruses are that they cannot grow outside a living cell and that, for many of them, the range of cell types in which they will replicate is limited. A few viruses cannot be grown in the laboratory at all, suggesting that a cell growing in a culture vessel ('*in vitro*') is rather different from the same type of cell in its normal complex environment within the body ('*in vivo*'). Likewise, a virus growing in a cell culture may be rather different from a similar virus replicating in nature. This fundamental problem is always with virologists and is particularly relevant to the matching of a strain used in a vaccine with the natural virus. It is less of a difficulty for diagnostic work, when usually all we want to know is that a given virus has been isolated.

As long ago as the 1930s Goodpasture, in a classic series of investigations, pioneered the use of the **chick embryo** to grow viruses. In theory, this was a most unlikely substrate in which to grow mammalian viruses, but it nevertheless proved very useful for propagating many of them, including influenza, herpes-, paramyxo-, and poxviruses. Even now, when a new virus such as HIV appears, one of the first experiments is to inoculate eggs by a variety of routes to see if it replicates. But over the last three decades the use of chick embryos for primary virus isolation has declined and most laboratories now use mammalian cell cultures for primary isolation. However, in view both of their historical importance and of their present-day use for making influenza vaccines, we shall briefly mention the main procedures without going into technical details.

True cell culture methods developed slowly until the advent of antibiotics greatly diminished the risk of bacterial contamination and permitted the handling of cultures on the open bench. Thereafter, almost every human and animal tissue was investigated, either as explants, i.e. fragments of tissue embedded in plasma clots or, later, as layers of cells on glass or plastic surfaces. A particularly significant event was the discovery in 1957 by Enders and his colleagues that poliovirus could be grown in monkey kidney cell cultures. This discovery, for which they received the Nobel Prize for Medicine, enabled the production of polio vaccine and also broke new ground scientifically, since it showed that, *in vitro*, viruses could grow in cells from organs that they would not infect in the intact animal or human host.

2 PROPAGATION IN CELL AND TISSUE CULTURES

These terms are often used synonymously; **cell culture** should, however, be reserved for the propagation of dispersed cells, either in suspensions or as continuous (confluent) sheets adhering to glass or plastic surfaces. Such sheets, one cell thick, are called **monolayers**. **Tissue culture** refers to the growth of fragments of unorganized tissue, usually fibroblasts, in plasma clots; this method is now rarely used and then only for research purposes. **Organ culture** is the maintenance *in vitro* of pieces of organized tissue, e.g. trachea. In the diagnostic laboratory, cell culture is the routine method.

2.1 *Culture media*

The **growth medium** used to cultivate cells or tissues contains a solution of salts at physiological concentrations, glucose, amino acids, essential vitamins, and antibiotics; it is buffered at pH 7.2–7.4. Fetal calf serum is added to a concentration of 10–20 per cent to provide supplements essential for cell growth. When the cells have grown sufficiently for inoculation, the growth medium is replaced by a **maintenance medium** containing only 2–5 per cent serum, which permits little or no further multiplication.

2.2 *Primary cell cultures*

Fetal or adult tissue is collected aseptically and chopped into small pieces about 2 mm^3. Incubation with trypsin for 30 min reduces most of the tissue to a suspension of individual cells or small clumps, which are centrifuged to remove excess trypsin, suspended in culture medium, and incubated at 37°C in screw-capped glass or plastic tubes. During the next few days the cells form a continuous (confluent) layer. The culture tubes are often incubated on their sides in a slow roller apparatus, which exposes the cells alternately to the gas and liquid phases within the tube.

Such cultures are at first a mixture of fibroblastic, epithelial, and other cells, but the faster dividing fibroblasts tend to outgrow the others. For diagnostic work, primary cultures have been largely replaced by continuous or semicontinuous cell lines. They do, however, have the advantage that, being diploid, they can be used for virus vaccine production. For example, rabbit kidney cells are used for rubella vaccine, chick fibroblasts for measles vaccine, and monkey kidney cells for polio vaccine.

2.3 Cell line cultures

Semicontinuous cell lines

Such cells are derived from human or animal fetal tissue (Table 3.1). They have the normal **diploid** karyotype and hence can be used for vaccine production. Some lines are also used for diagnostic work.

These semicontinuous cells have a limited lifespan and can be subcultured through only 50 or so generations. Even so, millions of cultures can be obtained from a single fetal organ. In practice, several thousand vials of cells suspended in dimethyl sulphoxide are frozen in liquid nitrogen after only a few passages in the laboratory. Each of these can be used to generate further cultures (the 'seed lot' system). For vaccine production, only cells at a given low passage level, e.g. the fourth, are used and each batch of cells is carefully tested for karyotype and other properties. Such strict criteria are not necessary in the diagnostic laboratory, where cells from several successive subcultures may be used for virus isolation.

Table 3.1 Some cell lines used for propagating viruses

	Cells derived from
Semicontinuous (diploid) cell lines	
HDCS	Human fetal lung
MRC-9	
WI-38	
Continuous (aneuploid) cell lines	
HeLa	Human cervical carcinoma
Hep-2	Human epithelium
Vero	Monkey kidney
MDCK	Dog kidney

Continuous cell lines

These are the most widely used for diagnostic work (Table 3.1). They are derived either from a **tumour** or from normal cells, which, after repeated culture, have become **transformed** so that they behave like tumour-derived cells, i.e. they can be propagated indefinitely. They are **aneuploid**, i.e. they have an abnormal number of chromosomes. Because of their malignant characteristics (Chapter 6), such cells cannot be used to prepare vaccines for use in humans.

Lymphocyte cultures

Normal mature lymphocytes cannot be propagated in culture without special treatments. **B lymphocytes** (Chapter 5) will however divide and continue to do so indefinitely if infected with **Epstein–Barr virus** (EBV, Chapter 15). This characteristic, known as 'immortalization' is an example of cell transformation by a virus (Chapter 6) and can be used as a marker of isolation of EBV.

T lymphocytes will grow in the presence of a lymphokine, **interleukin-2 (IL-2)**, formerly known as T-cell growth factor. This finding proved essential to the study of the human retroviruses (HIV and HTLV, Chapter 22), which can be propagated in such cultures with the formation of syncytial giant cells.

Propagation of cell lines

Monolayers. A monolayer in a culture flask is treated with trypsin or versene to disperse it into a suspension of individual cells, which are then diluted in growth medium to a concentration of 10^5 to 10^6 per ml, and distributed into other flasks, tubes, or petri dishes for further subculture.

Usually, for virus isolation, small stoppered test tubes are used, incubated at $37°C$ at a slight slope. Within an hour the cells attach to the side of the tube and begin to divide, to give a confluent monolayer by 48 h. The growth medium is then substituted by maintenance medium and the tubes are inoculated with a small volume of the clinical specimen, e.g. throat washing, stool suspension, or vesicle fluid.

Mass cultures. For propagating large quantities of virus, e.g. for vaccine production, it is necessary to grow very large numbers of cells. Some cells can be grown continuously in **suspension** in fermenters containing thousands of litres, within which the composition of the medium is kept constant automatically (Fig. 3.1). This technique is used for making therapeutic interferon (Chapter 29). For cells that will grow only on **surfaces**, there are ingenious means of creating very large surface areas within small spaces. For example, some cells can be grown on Sephadex beads, about 0.2 mm in diameter, or on plastic sheet spirally wound within a cylinder.

Fig. 3.1. Cultivation of virus-infected cells on industrial scale for vaccine production. (By courtesy of the Public Health Laboratory Service Centre for Applied Microbiology and Research, Porton Down.)

Cytopathic effects

After adding virus the cell sheet is observed daily for a **cytopathic effect (CPE)** (Fig. 3.2). These changes in cell morphology are often characteristic of particular viruses (Table 3.2) and give an indication of the type that has been isolated, although this must often be confirmed by further tests. We shall describe cytopathic effects in more detail when we discuss diagnostic tests in Chapter 27; some photographs are shown in Fig. 27.4. Certain viruses induce little or nothing in the way of CPE and their presence must be detected by other means (Table 3.2).

Another method used increasingly often is **immunofluorescence**, in which virus or viral antigen within the cells is stained with a specific antiserum conjugated to a fluorescent dye and viewed under an ultraviolet microscope (Chapter 27). This technique has the advantages of **rapidity**, in that slow-growing agents such as cytomegalovirus can be detected well before they produce CPE, and **specificity**, since a positive result automatically identifies the virus.

Plaque formation

Each infective virus particle gives rise to one focus of infected cells. These individual foci of CPE, if unchecked, tend with successive cycles of viral replication to spread throughout the monolayer. By covering the latter with a layer of agarose or soft agar directly after inoculation, spread of CPE is limited to the cells immediately adjacent to those first infected. A suitably diluted virus suspension thus gives rise to a number of circular cleared areas

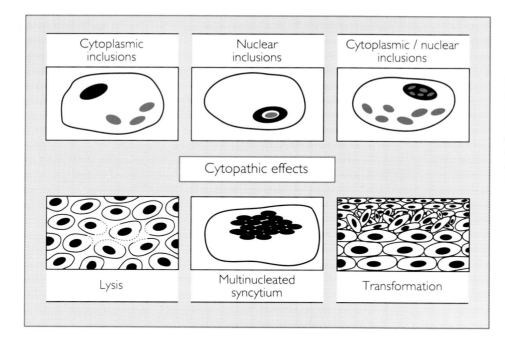

Fig. 3.2. Cytopathic effects of viral infection. (*top*) inclusion bodies: intracytoplasmic, e.g. rabies; intranuclear, e.g. herpesviruses; or both, e.g. measles virus. (*bottom*) the three main types of cytopathic effect: lysis, syncytium formation, and transformation.

Table 3.2 Effects of viruses on cell cultures

Virus	Effect
Cytopathic effects	
Picornaviruses	Rounding of cells; destruction of monolayer
Herpes simplex	Ballooning of cells, more pronounced with type 2 than type 1 virus
Adenoviruses	Clumps of rounded cells like bunches of grapes
Paramyxoviruses, human immunodeficiency virus (HIV)	Multinucleate giant cells (syncytia)
Non-cytopathic effects	
Orthomyxoviruses, paramyxoviruses	Render cells capable of adsorbing erythrocytes (haemadsorption)
Rubella, rhinoviruses	Interfere with and prevent cytopathic effect of another virus, e.g. an ECHO virus

within the monolayer, known as **plaques** (Fig. 3.3). These are exactly analogous to those caused in lawns of bacteria by bacteriophages and, like them, can be counted to estimate the number of infective viruses in the original suspension. Plaques caused by both bacterial and animal viruses have another important use, this time in genetic studies, in which those generated by mutated strains can be recognized by alterations in their appearance.

2.4 *Organ cultures*

Whereas cell or tissue cultures consist only of one or perhaps two kinds of cell, organ cultures are more complex. Some viruses will only grow in such a substrate, which closely mimics their original environment. For example, certain fastidious viruses of the respiratory tract, e.g. coronaviruses, grow only in cultures of human trachea, which are complete with beating cilia and mucus-producing cells.

Organ cultures of brain tissue, lung, intestine, or kidney, etc. are fragments of tissue about the size of a pea incubated in small petri dishes. In general, these cultures are difficult to maintain and are used only for research purposes.

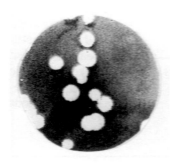

Fig. 3.3. Virus-induced plaques in a cell monolayer. The cells were infected with influenza virus and overlaid with agar. After incubation for a week, the preparation was stained. The clear areas in which the cells have been destroyed are the plaques, each of which is a focus of infection initiated by a single infective virion.

3 PROPAGATION IN CHICK EMBRYOS

The time from fertilization of hens' eggs to hatching is 21 days. Before inoculation, the eggs are incubated at 37°C, usually for 10–14 days, the exact period depending on the route, which varies with the virus under investigation.

Amniotic inoculation of 10–11 day old embryos is commonly used to isolate influenza viruses. A small volume of throat washing is inoculated; the virus, if present, replicates in the embryonic lung cells and is harvested 2 or 3 days later by sucking off the amniotic fluid. Antibiotics are added to the specimen to suppress growth of unwanted bacteria.

Allantoic inoculation. Influenza viruses adapted to chick embryo tissue by growth in the amnion can be propagated in much larger quantities within the allantoic cavity, the fluid from which is collected 48 h after inoculation. This method is used for vaccine production, harvesting of the allantoic fluid (about 5 ml per egg) being automated.

Chorioallantoic membrane. On this membrane, pox and herpes simplex viruses produce discrete lesions ('pocks') 1–3 mm in diameter, each of which is a focus of cell proliferation generated by the replication of one virus particle. Pock counts were used to measure the infectivity of suspensions of these viruses, but the technique has been superseded by cell culture methods.

4 PROPAGATION IN ANIMALS

The first virus vaccines, smallpox and rabies, were prepared in animals, the one on the skin of calves, sheep, or water-buffaloes, the other in the spinal

HOW VIRUSES CAUSE DISEASE

1 INTRODUCTION

Figure 4.1 shows a child with measles, an elderly patient with severe herpes zoster ('shingles'), and another child with a malignant tumour of the jaw known as Burkitt's lymphoma. Each of these people is under attack by a virus, but the manifestations differ greatly, not only in appearance but also in the ways in which the viruses concerned caused these unpleasant effects. This chapter is concerned with the complex interactions between viruses and hosts which result in disease, in other words, with the **pathogenesis** of viral infection.

2 VIRAL FACTORS: PATHOGENICITY AND VIRULENCE

The terms **pathogenicity** and **virulence** are often used interchangeably, but this is not strictly correct. In brief, 'pathogenicity' compares the severity of disease caused by **different micro-organisms**: rabies virus is more pathogenic than measles. 'Virulence', on the other hand, compares the severity of the disease caused by **different strains of the same micro-organism**. In practice this may be related to the numbers of organisms needed to produce a given effect, e.g. the death of a mouse. For example, two strains of herpes simplex virus inoculated into the skin may cause vesicular lesions, and are

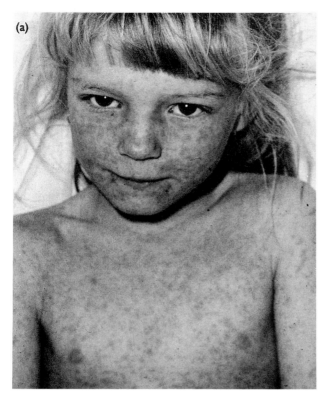

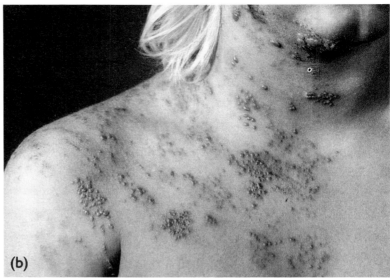

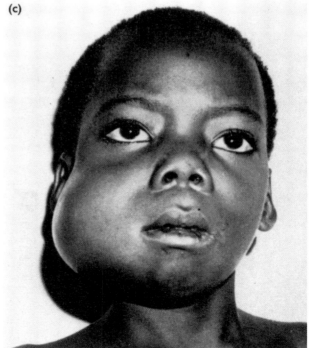

Fig. 4.1. Differing manifestations of viral infections. (a) Child with measles (by courtesy of the late Dr W. Marshall). (b) Elderly patient with herpes zoster. (c) Seven-year-old boy with Burkitt's lymphoma involving the right mandible (by courtesy of Dr Joan Edwards).

thus both pathogenic; but as few as 10 virions of strain A may kill the mouse, whereas 10 000 virions of strain B are needed to do so. Strain A is thus a thousand times more virulent than strain B.

What is it that makes one strain of virus more virulent than another? The development of rapid methods for determining nucleotide sequences within DNA and RNA are helping virologists to answer this intriguing question, which is important both from the point of view of preparing attenuated viruses for vaccines and of trying to predict whether viruses of enhanced virulence could suddenly be created as the result of a small mutational change. The emergence of AIDS might well be an example of some such event.

In at least some RNA viruses only very few nucleotide substitutions in the viral genome are needed to control a switch from virulence to avirulence. Poliomyelitis and influenza viruses provide two excellent examples. The complete genomes of a virulent type III poliovirus and an attenuated strain derived from it have been sequenced and compared. Only 10 **point mutations** — i.e. changes in single nucleotides — out of a total of about 7430 bases were detected in the attenuated strain; it now appears that only three of them give rise to amino-acid substitutions and, of these, only one point mutation may be responsible for attenuation of virulence. Likewise, a single change in amino-acid sequence near the receptor-binding site at the top of the HA molecule had a decisive influence on the virulence of influenza B for volunteers (Fig. 4.2).

The finding that virulence or lack of it depends upon such small molecular changes in the genome was unexpected and opens up new vistas both for vaccine development and for epidemiology.

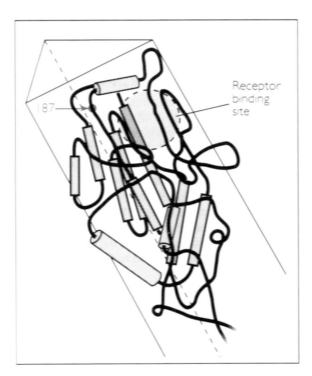

Receptor binding site

Fig. 4.2. X-ray crystallographic structure of influenza haemagglutinin. Arrow indicates where substitution of a single amino-acid (no.187) causes a change in virulence.

To cause disease, a virus has to clear a number of hurdles that vary somewhat in type and number according both to the virus concerned and its host. The following sequence of events is typical. The virus must:

1. **invade** the host;
2. **establish a bridgehead** by replicating in susceptible cells at the site of inoculation;

3 INTERACTIONS BETWEEN VIRUSES AND HOST CELLS

The interactions between the virus and the cells within which it replicates are of decisive importance in determining whether infection takes place at all, the

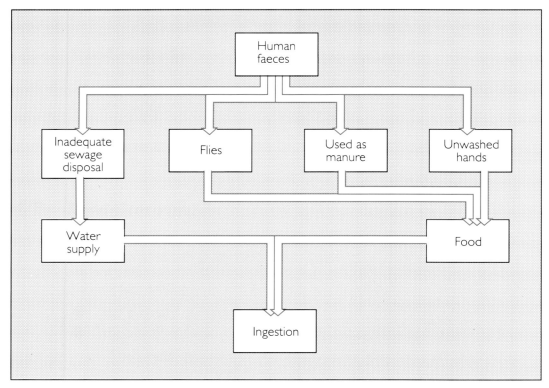

Fig. 4.3. The faecal–oral route of transmission of viruses.

revolting and is, since it means that viruses shed in faeces have got into someone else's mouth. Figure 4.3 shows a number of ways in which this can happen. Such infections are by no means confined to Third World countries; they are prevalent wherever sanitary conditions are indifferent, for example, in some mental institutions.

The incidence of **sexually transmitted diseases (STD)** has increased enormously since the end of the Second World War in 1945. Much of the increase is due to viruses and chlamydia; the 'classical' venereal infections, gonorrhoea and syphilis, are more easily diagnosed and treated. Although some syndromes such as AIDS and hepatitis B are in Western Countries particularly associated with promiscuous male homosexuals, all the STD mentioned in Table 4.2 can also be spread by heterosexual intercourse.

Another and increasingly frequent way of acquiring infection (not mentioned in the tables) is from **transplants,** particularly of bone marrow and kidney. Two herpesviruses—cytomegalovirus and Epstein-Barr virus— are notorious in this respect. Both cause persistent but asymptomatic infections in a substantial proportion of the general population; if an organ donor is infected in this way, the recipient is liable to suffer, particularly if he or she has no pre-existing immunity or if the immune responses have been impaired by immunosuppressive treatment (Chapter 26). **Blood transfusions** and **blood products** such as Factor VIII are also a potential source of infection with certain viruses, such as hepatitis B, hepatitis C, and the AIDS virus, HIV.

Rigorous screening and the exclusion of high-risk donors such as drug-addicts has however greatly reduced — but not eliminated — the number of infections from such sources.

Other infections resulting from **surgical treatment** have been described, but are rare. More than one person has died from rabies after receiving a corneal transplant from a donor who was incubating the disease — surely the ultimate in bad luck stories! Others again have acquired Creutzfeldt–Jakob disease, another fatal infection of the central nervous system, from surgical instruments contaminated with the causal agent, which is unusually resistant to sterilization (Chapter 23).

Spread of infection from mother to fetus is a special form of transmission from one person to another and is described in Chapter 25.

Like bacterial and fungal infections, those caused by viruses can be classified under two main headings:

◆   those localized to tissues at or contiguous with the site of entry;

◆   those that spread to one or more organs remote from this area.

4.3 *Localized infections*

These are infections of epithelial surfaces: the skin, the conjunctiva, and the mucous membranes of the respiratory, gastrointestinal, and genital tracts. Some examples are given in Tables 4.1 and 4.2.

Localized infections of the skin by poxviruses result in papular lesions, usually proceeding to vesicle and then pustule formation (e.g. vaccinia, orf), or in proliferative lesions of the epidermis (molluscum contagiosum and papillomaviruses).

By contrast, infections of **mucous membranes** spread over comparatively large areas of the respiratory or gastrointestinal epithelium. The process is rapid, which means·that such infections have a short incubation period, say 1–3 days. Although viral replication is restricted to these surfaces, the effects may be much more general, as anyone who has had influenza or even a bad cold knows only too well.

4.4 *Generalized infections*

For various reasons, the pathogenesis of many generalized virus infections is not well understood: in some instances there is no suitable animal model; in others, such as the viral haemorrhagic fevers, the difficulty of working with dangerous pathogens under restricted conditions discourages extensive experimentation; and, in others again, our lack of knowledge results from inability to grow the virus in the laboratory. Nevertheless, a number of viruses, notably those causing the infectious fevers of childhood, follow a pattern of spread that was worked out by F. Fenner, in Australia, who studied a poxvirus

Table 4.3 Typical pattern of replication and spread of virus in an acute generalized viral infection

Event	Location
Infection and initial **replication**	Respiratory or intestinal epithelium
Replication—first stage of amplification	Local lymphoid tissue—regional lymph nodes, Peyer's patches
Primary viraemia	Blood
Replication—second stage of amplification	Reticuloendothelial organs—liver, spleen, bone marrow
Secondary viraemia	Blood
Replication in target organs	Skin, viscera, CNS, etc.

infection of mice called ectromelia; the sequence of events, summarized in Table 4.3, is as follows.

(1) Virions enter through an epithelial surface where they undergo limited replication. (2) They then migrate to the **regional lymph nodes** where some are taken up by macrophages and inactivated, but others enter the bloodstream. This is (3) the **primary viraemia**, which sometimes gives rise to prodromal malaise and fever.

(4) From the blood, virus gains access to the large reticuloendothelial organs—**liver, spleen** and **bone marrow**—in which it again multiplies. In this further cycle of amplification, a large amount of virus is produced which again, in a manner of speaking, spills over into the bloodstream, causing (5) a **secondary viraemia**.

(6) From the bloodstream, virus finally reaches its **target organ**, the nature of which depends on the tissue tropism of the virus concerned, and which mainly determines the clinical features of the illness. All this takes time, which explains why the incubation periods of infections of this type are of the order of 2 weeks.

In other infections, viruses reach their target organs by more straight-forward routes, in which the virus is injected directly into the bloodstream. This may happen in, for example, certain arbovirus infections.

4.5   *Some important target organs*

The skin. A rash, or **exanthem**, features in a number of virus infections (Table 4.4). Dental students in particular should note that it is sometimes accompanied by an **enanthem**, a rash affecting mucous membranes, which is, of course, best observed on the buccal mucosa (Table 4.5.).

There are several types, which are not necessary exclusive to particular viruses. Some viruses, notably the entero- and ECHO viruses, are notorious for causing almost any type of non-haemorrhagic rash.

Table 4.4 Examples of virus diseases associated with a skin rash (exanthem)

Disease	Type of rash
Rubella, erythema infectiosum (parvovirus), entero- and ECHO virus infections, dengue	Macular
Measles, infectious mononucleosis, entero- and ECHO virus infections	Maculopapular
Herpes simplex, chickenpox/herpes zoster, poxvirus infections, Coxsackie virus infections (especially types A9 and A16), entero- and ECHO virus infections	Vesicular or vesiculopustular
Congenital rubella	Purpuric
Viral haemorrhagic fevers	Petechial/haemorrhagic

Table 4.5 Virus diseases associated with a rash (enanthem) on the buccal mucous membrane

Disease	Main features of the enanthem
Herpesvirus infections	
Herpes simplex (usually primary infection)	Gingivostomatitis, with vesicles and ulcers on an intensely inflamed mucosa, primarily affecting the **anterior** part of the buccal cavity
Chickenpox	Vesicles, rapidly ulcerating, especially on fauces and palate
Infectious mononucleosis	Petechiae occasionally seen on palate
Coxsackie A virus infections	
Herpangina	Small vesicles usually confined to **posterior** part of the buccal cavity (soft palate and fauces)
Hand, foot, and mouth disease	Vesicles and ulcers anywhere on buccal mucosa, but predominating in **anterior** part
Paramyxovirus infection	
Measles	Koplik's spots on congested mucosa near openings of parotid ducts; appear a day or two before the skin rash

Note. These and many other acute viral infections may be accompanied by pharyngitis.

Whereas the vesicular eruptions are due to replication of virus in the skin, with conseqent damage to epithelial cells, other rashes may have other causes. Thus the characteristic maculo-papular rash of measles is due to the destruction of infected cells by cytotoxic T lymphocytes, purpuric rashes are often associated with a fall in blood platelets, and some haemorrhagic rashes are the result of disseminated intravascular coagulation (DIC).

The lung. Most respiratory infections, even those of the lung, result from local spread of virus as described in Section 4.3. Sometimes, however, the lung is involved as part of a generalized infection. This is particularly so in measles, in which some degree of pneumonitis is a constant feature. In patients whose immunity is damaged, both measles and varicella viruses may cause giant-cell pneumonia, characterized by the appearance of syncytia (Section 3.2 and see Figs 26.2 and 26.3); this is a dreaded complication with a high mortality.

The liver. This organ is the target in hepatitis A, hepatitis B, and 'non-A non-B hepatitis' (Chapter 20). It may also be damaged as part of a more general infection of body tissues, the classical example being yellow fever. Other virus infections come into this category and are listed in Table 20.1.

The kidney is rarely infected by viruses. The important exception is cytomegalovirus (CMV), one of the herpes group. Characteristic inclusions are found in the proximal renal tubules from which the virus is shed into the urine.

The central nervous system (CNS). Viruses gain access in one of two ways:

◆ From the **bloodstream**, during an episode of viraemia, e.g. poliomyelitis, arbovirus encephalitis;

◆ via the **nerves** connecting the periphery with various parts of the CNS, e.g. herpes simplex, varicella-zoster, rabies.

4.6   *Immunopathological damage*

We have mentioned that some skin rashes are manifestations of the immune response rather than the result of tissue damage by virus multiplication. This is often true of pathological changes in other organs infected by viruses (Chapter 5).

4.7   *Incubation periods*

Knowledge of the incubation periods of the common virus infections is important not only as an aid to the diagnosis of individual patients, but also as an essential tool in tracing the spread of outbreaks (Chapter 7). They will be given in more detail in the chapters devoted to individual viruses, but for the

Table 4.6 Incubation periods of representative viral diseases

Disease	Usual period	Time range
Short incubation (times in days)		
Enterovirus conjunctivitis	1–2	
Common cold	1–3	
Influenza	1–3	
Arbovirus infections	3–6	2–15
Medium incubation (times in days)		
Poliomyelitis	7–14	2–35
Measles	13–14	8–14
Rubella	14–16	14–21
Varicella	13–17	11–21
Mumps	14–18	14–21
Long incubation (times in weeks)		
Hepatitis A	3–5	2–6
Hepatitis B	10–12	6–20
Infectious mononucleosis	4–6	2–7
Rabies	4–7	2–50
Very long incubation		
Subacute sclerosing panencephalitis (SSPE)		
Progressive multifocal leucoencephalopathy	Years	
Creutzfeld–Jakob disease		

moment, and as an aid to memory, we shall classify them into four main groups: short, medium, long, and very long (Table 4.6).

Short means less than a week and primarily applies to viruses causing localized infections that spread rapidly on mucous surfaces. Some viruses injected directly into the blood, e.g. arboviruses transmitted by the bite of an arthropod, also as a rule have short incubation periods.

Medium incubation periods range from about 7 to 21 days; they are seen in generalized infections having the type of pathogenesis described in Section 4.4.

Long refers to periods measured in weeks or months, e.g. 2–6 weeks for hepatitis A and 6–20 weeks for hepatitis B. The pathogenesis of these infections has not yet been worked out, and we do not know what these viruses are doing between entering the host and producing symptoms and signs of illness. Rabies may also have incubation periods extending for many months, but in this instance we know that the time to onset depends on the time taken for the virus to travel up neurons to the brain from the site of entry, and thus on the length of the nerves affected.

Very long incubation periods are measured in years, which is why some of the agents involved are known as 'slow' viruses. This group comprises the 'unconventional' virus-like agents, whose nature is not well understood (Chapter 23); and papovaviruses and measles, which very occasionally cause delayed disease of the central nervous system. These infections are invariably fatal. Why their incubation periods are so long remains a mystery.

Table 4.7 Latent virus infections

Virus	Site of latency
Herpesviruses	
Herpes simplex types 1 and 2; varicella-zoster	Neurons in dorsal root ganglia
Epstein–Barr virus	B lymphocytes
Cytomegalovirus	Lymphocytes; macrophages (?)
Human herpes virus type 6	Probably lymphocytes
Hepadnavirus	
Hepatitis B	Hepatocytes
Papovavirus	
Papillomavirus	Epithelium
Retroviruses	
Endogenous oncornaviruses	Somatic and germ cells
Lentiviruses	
Human immunodeficiency viruses	T lymphocytes, macrophages, brain cells

virus without being destroyed. Such infections have their counterparts *in vivo* (group 3B). Unlike latent infections, they may be caused both by DNA and RNA viruses and the end result for the host is often determined by an abnormal or defective response of the immune system, as may be the case in subacute sclerosing panencephalitis (SSPE), a fatal but fortunately very rare late complication of measles; chronic active hepatitis B is another example. Other chronic but more trivial infections are caused by the papillomaviruses responsible for warts and by the poxvirus which gives rise to molluscum contagiosum. In both these skin conditions, the viruses evade the immune response by remaining within their safe haven in the avascular epidermis. Papillomavirus genome may also integrate with host-cell DNA.

5.3 *Insidious infections with fatal outcomes*
(Fig. 4.4, group 3)

The two types of infection in this group resemble each other only superficially. The 'slow' virus infections (group 3A) have already been mentioned in Section 4.7. The other type of infection (group 3B) is not known to occur in man, but is of great interest both to virologists and to immunologists. Indeed, it was the study of lymphocytic choriomeningitis, an arenavirus infection of mice, that led the late Sir Macfarlane Burnet to the idea of **immunological tolerance** and a Nobel Prize. In brief, every member of a mouse colony in which this virus is present becomes infected at birth and continues so for the rest of its life. In such mice, the immune system fails to recognize the virus-infected cells as 'foreign'. Some antibody is produced, but forms virus–antibody complexes; these are not dealt with in the normal way, and a proportion of infected mice eventually die because of their deposition in the kidney.

6 SHEDDING OF VIRUS FROM THE HOST

Microbes only live to fight another day by getting out of one host and into another; the means of escape is therefore very important. Viruses may be shed from the **primary site of multiplication** or, in the case of generalized infections, from the **target organ**. Viruses can readily escape in one way or another from all the main body systems, with the exception of the CNS, in which they are effectively bottled up. It is important to remember that many viruses may be shed from clinically normal people: they include herpes simplex (saliva), cytomegalovirus (urine, breast milk), and the viruses that infect the gut (faeces).

We shall see in Chapter 7 what happens when viruses escape into the community.

7 REMINDERS

◆ Viruses gain access to the host through **skin** and **mucous membranes**, and via the **respiratory, gastrointestinal**, and **genital tracts**.

◆ Their virulence or lack of it may be determined by very small mutations in the genome.

◆ Viral infection of cells may result in **lysis, fusion** and **syncytium formation**, the appearance of **inclusion bodies**, or, in some instances, **transformation** to cells with characteristics of malignancy.

◆ Infections caused by viruses are **localized** at or near the site of entry or **generalized** to involve one or more **target organs**. In generalized infections a common pattern of spread during the incubation period is **site of entry→local lymph nodes→primary viraemia→liver, spleen, bone marrow→secondary viraemia→target organ**.

◆ In humans, viral infections follow various basic patterns: **acute non-persistent, acute followed by persistent latent infection**, and **chronic with continued shedding of virus**. A group of unconventional viruses causes long-lasting and fatal '**slow**' infections.

◆ During the course of infection, viruses are **shed** from the primary site of multiplication, or — with the exception of the central nervous system — from target organs, and are then free to infect other susceptible hosts.

RESISTANCE TO INFECTION

1 INTRODUCTION

So far, the emphasis has been almost entirely on the ingenious means adopted by viruses to invade and damage their hosts. On its part, however, the animal world has developed efficient and highly complex mechanisms for combating infection by parasites, which are usually considered under the headings of **general** (or 'non-specific') and **specific** resistance. The latter term refers to the array of immune responses directed against particular pathogens, but it is not always possible to make a hard and fast distinction between this type of resistance and the more general defences, which will be dealt with first.

2 GENERAL FACTORS IN RESISTANCE

These fall into two categories: those that protect the individual (Sections 2.1–2.5) and others, genetically mediated, that determine the resistance or susceptibility of populations (Sections 2.6 and 2.7).

2.1 *Mechanical and chemical barriers*

The importance of the skin as a barrier to infection was mentioned in Chapter 4, Section 4.2. The retrograde movement of epithelial cilia acts to prevent infection of the respiratory tract; damage to these cells by influenza and paramyxoviruses may open the way to secondary bacterial infection. The gastrointestinal tract is to some extent protected against ingested viruses by the low pH in the stomach, although viruses that regularly infect by this route are resistant to acidity; this applies to most enteroviruses, but it is interesting that the rhinoviruses, a subgroup that cause the common cold, do not have to survive passage through the stomach and are not acid-resistant.

2.2 *Fever*

A high temperature is naturally regarded by patients as an unpleasant effect of the virus infection; this response of the body should, however, be welcomed since it is in fact a defence mechanism. It is now recognized that temperatures much above 37°C are inimical to the replication of a number of viruses, and that the taking of antipyretic drugs such as aspirin may delay rather than help recovery. The febrile response seems to be triggered by soluble factors, notably interleukin-1 (produced by macrophages) and by interferons.

2.3 *Age*

This factor is an example of the way in which the general overlaps with the particular, since age-related resistance is, in part at least, mediated by immune responses. An infant is sent into the world with a useful leaving present from its mother in the form of a package of IgG antibodies directed against infections from which she has suffered. Immunity to these predominantly viral infections, supplemented by IgA antibodies in colostrum and breast milk, helps to tide the baby over the first 6 months or so, after which its susceptibility to viral infections increases. The protection conferred by maternal milk is a good reason for breast-feeding, especially in Third World countries where the energetic sale of manufactured substitutes is not in the best interest of babies born into a particularly hostile environment. Another great advantage of breast milk is that it is bacteriologically sterile. The United Nations Children's Fund (UNICEF) estimates that the lives of 1.5 million infants—mostly in developing countries—could be saved by halting the decline in breast feeding.

Virus infections in childhood are not usually serious, but become increasingly so with the advance of age; for example, poliomyelitis usually causes mild or even subclinical infection in children, but adults are often hit harder and in them the incidence of paralytic disease is higher. Virus infections in elderly people can be very severe, herpes zoster (shingles) being a case in point; such increased susceptibility may be due at least in part to failure of a tired immune system.

2.4 *Nutritional status*

Poor nutrition may exacerbate the severity of some virus infections, the most often-quoted example being measles in African children, which has a much higher mortality rate than in developed countries, but assessment of the importance of malnutrition is complicated by other factors such as intercurrent infections, particularly malaria, which is immunosuppressive.

2.5 *Hormones*

It is well known that treatment with steroids increases the severity of herpes simplex and varicella-zoster infections but their precise role in natural resistance or susceptibility is unknown. The severity of viral hepatitis is sometimes exacerbated by pregnancy, presumably because of hormonal influences, but again, the mechanism is not yet understood.

2.6 *Genetic factors*

In experimental animals there is clear evidence that genetic factors influence resistance or, conversely, susceptibility to virus infections. Thus, some highly inbred lines of mice are killed by very small inocula of herpes simplex virus, whereas others withstand enormous doses with no sign of illness. In this case resistance is dominant and is mediated by only four genes; with other viruses, susceptibility may be the dominant genetic factor. The genes involved are sometimes, but not always, part of the major histocompatibility complex (MHC).

   The influence of genetic factors can to some extent be determined by these neat and tidy experiments involving inbred animals and well-defined virus strains. Needless to say, the assessment of such factors in humans is at present far more difficult, but is being helped by continued progress in our understanding of the genes that mediate immune responses.

2.7 *Species resistance*

The host range of many viruses is restricted, probably because the cells of resistant species do not possess appropriate receptors. The best understood example is poliovirus, the receptors for which are present only in man and other primates. Others, notably the human immunodeficiency viruses and some hepatitis agents, are equally selective; by contrast, others again, such as rabies, are capable of infecting most or all warm-blooded animals.

3 THE IMMUNE RESPONSE TO VIRUS INFECTIONS

3.1 *Introduction*

It was a virus infection, smallpox, that gave rise to the very first notion of immunity. During the eighteenth century, European visitors to China and India brought back tales of how material from the skin lesions of patients was used to inoculate healthy people; the resulting mild infection protected against the much more severe natural disease. An understanding of the immune responses to virus infections is important for several reasons.

First, the relationship between viruses and the immune system is much more intimate than it is for most bacteria. Viruses often modify the cells within which they replicate, thereby rendering them 'foreign' and susceptible to attack by sensitized lymphocytes; furthermore, some viruses can multiply within the very lymphocytes and mononuclear phagocytes that are important components of the immunological defences: a good example is the human immunodeficiency virus (HIV) which causes AIDS. There are other very practical reasons for knowing about these immune responses. By contrast with the larger independent parasites such as bacteria, fungi, and protozoa:

◆ Laboratory detection of viruses is comparatively difficult and serological methods of **diagnosis** are thus more important.

◆ Chemotherapy is usually unavailable, so that virus infections are largely combated by **immunization**. When considering this subject it is important to distinguish between **resistance to** and **recovery from** infection, since the mechanisms are often very different.

3.2 *The immune system*

Outline

Scientific investigation of immunology started about a century ago with Metchnikoff's studies of phagocytosis of foreign particles, including micro-organisms. This early interest in what we now refer to as cell-mediated immunity was soon overtaken by researches on antibody-mediated (humoral) immunity which was for long regarded as the primary defence against microbial disease. We now know that, although the presence of antibody is important in **preventing** virus infections, cellular immunity plays a major role in the **recovery** process.

The essential characteristics of the specific immune responses are as follows.

◆ Ability to distinguish between 'self' macromolecules that properly belong to the body and are ignored; and 'non-self' molecules, e.g. those of microbes. The latter are **antigens**, that is, they provoke a **specific** immune response.

◆ '**Memory**', whereby an enhanced response to an antigen previously encountered is evoked. Specific immunity is thus also referred to as **acquired** immunity, as opposed to the non-specific mechanisms with which we are born and which were described in the first part of this chapter.

Let us now look at the main components of the specific immune response.

Although antigens are usually large molecules such as proteins, only small segments of, say, half a dozen amino acids actually induce the formation of antibodies and react with them. The antigens of importance in immune responses to viruses are often **glycosylated** to form glycoproteins.

Immunological membory is mediated by **lymphocytes**, each of which is capable of responding to a single antigen specific for that particular cell. It is astonishing that during the course of evolution we have developed a range of lymphocytes that can react to virtually any one of the myriads of molecules capable of evoking an immune response: there are two main classes of lymphocyte, B cells and T cells, both of which react with antigens by means of **cell-surface receptors**. Although respectively responsible for antibody- and cell-mediated immunity, these two classes of lymphocyte should not be thought of as occupying separate compartments of the immune response. On the contrary, they co-operate closely with each other and with dendritic cells and macrophages that process antigens and 'present' them to the lymphocytes.

B cells and antibody-mediated immunity

The role of **B lymphocytes** is summarized in Fig. 5.1 (numbers in this section refer to steps in this figure). Note that in this figure and in Fig. 5.4, the nuclei of B and T lymphocytes are respectively stippled lightly and heavily. They originate from stem cells in the **bone marrow**, enter the circulation, and mature in the liver, spleen, and lymphoid tissue. Their surface receptors are **antibody molecules** that react on encountering their specific antigen, which usually has to be processed (1) by **antigen-presenting cells (APC)**. These are **dendritic cells** present in the germinal centres of the lymph nodes and spleen; along with mononuclear phagocytes (macrophages and monocytes), they process foreign antigenic material in a way not fully understood, and present it on their surfaces in a form that is recognized by a B lymphocyte bearing the corresponding antibody receptors (2). This encounter stimulates the B cell to proliferate, thus forming a **clone** of cells with the same antigenic specificity (3). The process is assisted by lymphokines secreted by T-helper (T_h) cells (4), which themselves are stimulated by reacting with the altered antigen, this time in association with class II histocompatibility antigens on the surface of the APC (5).

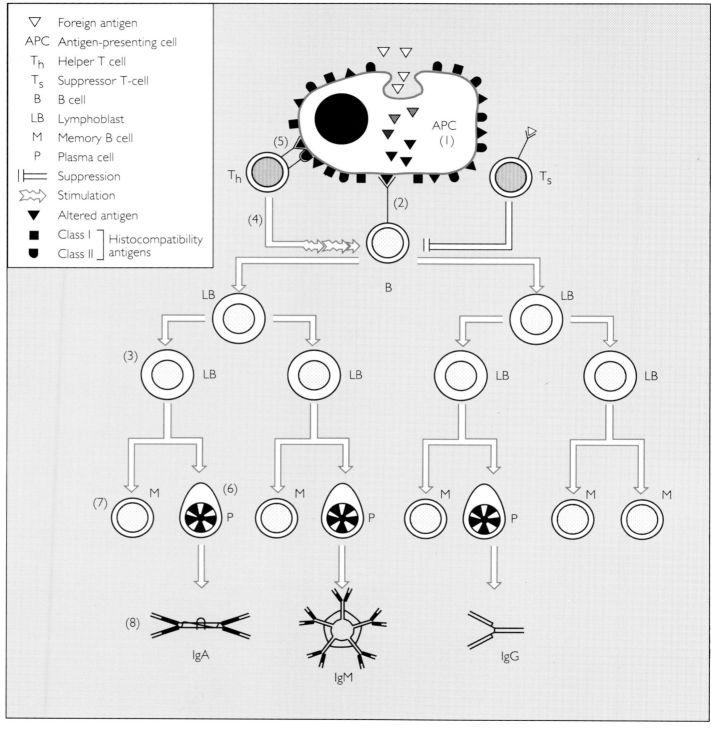

Fig. 5.1. The B-cell response to viral infection. For explanation see text.

Many of the cells in the clone differentiate into **effector cells**, in this instance **plasma cells** (6) that secrete more antibodies capable of combining with that antigen and that antigen only. Some of the sensitized B lymphocytes persist for long periods as **memory cells** (7) that rapidly respond to further encounters with the antigen by clonal proliferation and production of more of the appropriate antibody. The amount of antibody generated is controlled by T cells, of which more later.

The five classes of immunoglobulin, IgA, IgM, IgG, IgD, and IgE, are each produced by particular clones of plasma cells; only the first three seem to be important in virus infections (Table 5.1). Each consists of one or more units made up of four polypeptide chains, two 'heavy' and two 'light', which together form a Y-shaped molecule (Fig. 5.2). The tips of the two prongs of the 'Y' are the **variable regions**. Within the population of B lymphocytes, the extremely wide variation of amino-acid sequences in these regions ensures that there will be a few cells capable of reacting with any antigen encountered for the first time. The variable regions thus confer upon each immunoglobulin molecule its individual **antigenic specificity**.

IgA antibodies are produced by lymphoid tissue underlying the mucous membranes at whose surfaces it acts; they are found in secretions of the oropharynx and of the gastrointestinal, respiratory, and genital tracts, and are thus important in defending against viruses that enter by these routes (Section 2.1). The IgA secreted at mucous surfaces consists of two immunoglobulin units (a dimer) attached to a 'secretory piece' that aids its passage through cells. IgA is also produced during lactation, particularly in the colostrum, and it is the specific antibodies provided in this form by the mother that help to protect against infections in early infancy (Section 2.3).

Table 5.1 Immunoglobulins involved in immunity to viral infections

Properties	Immunoglobulin		
	IgM	IgA	IgG
Molecular weight	900 000	385 000*	150 000
Shape of molecule			
No. of heavy/light chain units	5 (pentamer)	2* (dimer)	1 (monomer)
Fixes complement	Yes	No	Yes
Crosses placenta	No	No	Yes

* In dimeric (two-molecule) form, in which the chains are linked to a secretory component (see text). Other polymers may also be formed.

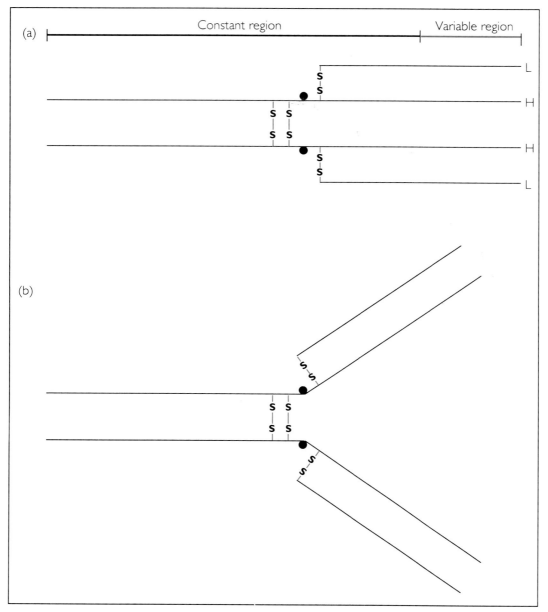

Fig. 5.2. Structure of an immunoglobulin molecule. (a) Basic structure. Two identical 'heavy' polypeptide chains (H) are linked to two identical 'light' chains (L) by disulphide bonds. (b) Molecule opens at hinge region. This is the position in which it combines with an antigen.

IgM antibodies are the first to be produced in systemic infections and are particularly avid in combining with antigen and complement. IgM is a large molecule and cannot cross the placenta; hence specific IgM antibody in a fetus or neonate indicates intrauterine infection. Production of IgM antibody is a fairly short-term process, lasting for a few weeks or months. **The finding of a specific IgM is thus evidence of a recent or current infection** and is now widely used for diagnosis.

Eventually, however, a negative feedback mechanism operates whereby the manufacture of IgM is switched off and replaced by production of specific IgG antibody.

IgG antibodies. By contrast with IgM, IgG antibodies continue to be produced for very long periods — often during the entire lifespan — and thus afford **long-term protection** against subsequent encounters with the same virus. Since their presence is evidence of past infection, they are useful in epidemiological surveys (Chapter 7). IgG antibodies cross the placenta and, as mentioned in Section 2.3, provide protection during the first few months of life, thus supplementing the action of IgA antibodies. The time-course of antibody production can be observed most clearly during immunization with a non-living antigen such as an inactivated vaccine (Fig. 5.3). The **primary response** to the first injection results from the initial clonal expansion of B lymphocytes on first encounter with the antigen; it is comparatively slow and of low magnitude. By contrast, the **secondary responses** to subsequent injections given some weeks later reflect the presence of sensitized memory cells that are now available to undergo a greatly amplified clonal expansion; the resultant IgG antibody response starts very quickly, i.e. within a day or so, and much greater quantities are produced. This clear-cut picture is blurred when a replicating antigen such as live polio vaccine is given, or during the course of a naturally acquired virus infection: in this instance the primary stimulus continues for some time and other immune mechanisms relating to recovery are brought into play.

Mode of action of antibodies

There are various ways in which a specific immunoglobulin can act against a virus.

◆   It can **neutralize** by combining with the virions; this may stop them attaching to susceptible cells or, more probably, interferes with the uncoating sequence or mRNA transcription after the virus has entered.

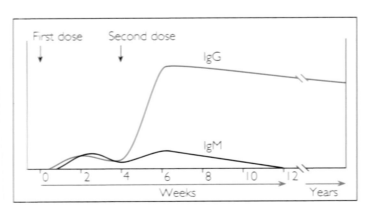

Fig. 5.3. Primary and secondary antibody responses to spaced doses of an inactivated vaccine.

◆ Antibody plus **complement** can combine with viral antigen expressed on the surface of an infected cell and lyse it. This effect is known as antibody-dependent cellular cytotoxicity (ADCC).

◆ Antibody may act as an **opsonin**, combining with virions and increasing the ability of macrophages to phagocytose and destroy them.

◆ **Macrophages** coated with specific antibody are **activated** or 'armed' to destroy infected cells expressing on their surfaces viral antigens with the same specificity.

T cells and cell-mediated immunity

Like B cells, **T lymphocytes** also originate from stem cells in the bone marrow, but then migrate to the **thymus** where they mature and acquire their specific antigen receptors. These cells are responsible for **cell-mediated immunity (CMI)**; there are various **subsets**, some of which operate by secreting soluble substances called **lymphokines** or **interleukins**. The main classes of T cell are now described.

Regulatory cells. These either promote or prevent the activities of other T cells and production of antibody by the B cells; they are known respectively as **T-helper (T_h)** and **T-suppressor (T_s)** cells, or, according to surface markers identified by monoclonal antibodies, as CD4 and CD8 (formerly T4 and T8). In the blood, their normal ratio is about 2 : 1. Most viral antigens are T-dependent, which means that B cells can make antibody to them only in co-operation with T lymphocytes.

Cytotoxic cells. These cells **(T_c)** are particularly important in virus infections, since they recognize virus-specified antigens on the surface of infected cells, which they attack and lyse.

Delayed hypersensitivity (T_{dh}) cells secrete lymphokines that activate other T cells and macrophages.

Like B cells, T lymphocytes undergo clonal expansion when appropriately stimulated, with the production of both effector and memory cells.

Mononuclear phagocytic cells (MP) comprise blood monocytes, fixed macrophages in the tissues (e.g. Kuppfer cells in the liver, microglia in the brain) and populations of free macrophages in the lungs, peritoneal cavity, and other tissues. These cells are real busybodies in the immune system: as well as acting as antigen-presenting cells and phagocytosing antigen–antibody complexes, they can lyse infected cells, produce interferon, and secrete other lymphokines that modulate the behaviour of T cells. In turn, their own behaviour is affected by activating or inhibiting factors produced by T cells.

MHC restriction. The operations of T cells are intimately related to the antigens of the major histocompatibility complex (MHC). Recognition of 'foreign' antigens takes place only if they are presented by the dendritic cell or macrophage in association with the correct MHC antigens: thus T_c and T_s cells act only when antigen is presented with class I MHC antigens identical with their own, whereas the activities of T_h lymphocytes are restricted by the need for homologous class II antigens on the presenting cell surface.

Now look again at Fig. 5.1. The B cell is being assisted in its functions by lymphokines secreted by the T helper cell on the left, which itself is activated by interaction with altered viral antigen on the surface of the antigen-presenting cell. Note particularly that this interaction can take place only in association with a class II MHC antigen normally present on the surface of the APC. On the right of the B lymphocyte, a T suppressor cell stands by ready to damp down overproduction of antibody by the B cell. This is the interface between antibody- and cell-mediated immunity.

Figure 5.4 illustrates some important interactions among T cells; numbers in the ensuing paragraphs refer to this figure. An antigen-presenting cell similar to that in Fig. 5.1 is shown; as a reminder, a B cell being helped by a T_h lymphocyte (1) is depicted at the left, but this time the centre of the stage is occupied by various T cells.

A cytotoxic T cell (T_c) is stimulated by reaction with viral antigen on the surface of the APC (2) to undergo clonal proliferation (not shown for clarity). In association with class I histocompatibility antigens, T_c cells attach to viral antigen on the surface of infected ('target') cells and lyse them (3). These activities are assisted by T_h cells acting through a soluble lymphokine, in this instance **interleukin-2** (4).

Another subset of T cells that mediates delayed hypersensitivity (T_{dh}) reacts with viral antigen on the surface of the APC in association either with class I or class II histocompatibility antigens (5); these too are assisted by T_h cells (6). The stimulated T_{dh} cells secrete chemotactic lymphokines that attract more T cells and some macrophages (7) to help in fire-fighting at the site of infection.

As in the B cell response, T_s cells stand by to moderate overly energetic behaviour by the T cells (8).

Figure 5.4 illustrates the central role of T lymphocytes in immunity to virus — and many other — infections, which has been likened to that of a conductor orchestrating the entire immune response.

Apart from mopping up antigen/antibody complexes, polymorphonuclear leukocytes are relatively unimportant in virus infections. This can readily be inferred from tissue sections or exudates in which, by contrast with bacterial infections, the overwhelming majority of cells are lymphocytes. An example is shown in Fig. 20.3.

Natural killer (NK) cells. In addition to T_c cells and macrophages there are lymphocytes of uncertain origin and without immunological specificity which can kill cells that they recognize as having 'foreign' antigens on their surfaces, i.e. virus-infected and cancer cells. Their activity is greatly enhanced by interferon, but their role in recovery from infection is still uncertain.

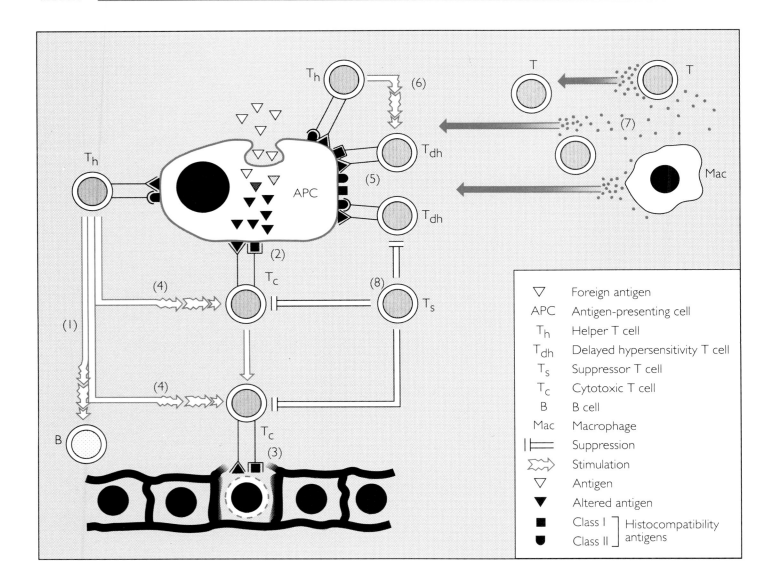

Interferon

The discovery by Isaacs and Lindenmann in 1957 that virus-infected cells produce a soluble factor that protects other cells from infection seemed to herald a new era, since the substance was effective against a wide range of viruses and apparently non-toxic. At a time when rapid strides were being made in the chemotherapy of bacterial infections, interferon (IFN) seemed like the answer to the virologists' prayers. These early hopes were soon to turn to disappointment, but now, with better understanding and technology, interferons are making something of a comeback.

Why the change to plural? It is now known that there are a number of interferons which differ in the way they are produced, in chemical composi-

Fig. 5.4. The T-cell response to viral infection. For explanation see text.

Table 5.2 Some properties of interferons

Property	Alpha*	Beta	Gamma
Stable at pH 2.0	Yes	Yes	No
Produced by	Leukocytes	Fibroblasts	T lymphocytes

* Alpha and beta interferons were together formerly referred to as type I and gamma interferon as type II. They are now often referred to as IFN–α, IFN–β, and IFN–γ.

tion, and in mode of action (Table 5.2). They are proteins with molecular weights of around 20 000, manufactured by leukocytes or fibroblasts in response not only to viral infection, but also to stimulation by natural or synthetic double-stranded RNA and some bacteria, e.g. chlamydia. **These molecules are not virus-specific** so that IFN induced by one virus is effective against others; on the other hand, they are in general **species-specific** so that IFN produced by, say, a guinea-pig, is ineffective in mouse or human cells. Gamma IFN differs in a major way from the alpha and beta varieties since it is produced by a subset of T memory cells in response to stimulation by an antigen previously encountered (although the IFN itself is still not virus-specific); it is thus regarded as a type of lymphokine.

IFNs do not kill viruses, nor do they act like antibodies. The mechanism is complex and indirect, but is summarized as simply as possible in Fig. 5.5; the numbered steps in the figure will now be described.

(1) Viruses attach by a specific receptor to a cell which is stimulated to produce interferon molecules. These diffuse out from the cell (2) and induce an **antiviral state** in neighbouring cells. First, IFN molecules attach to receptors on the nearby cells (3) and induce the formation of three enzymes, a **ribonuclease (RNase)**, a **protein kinase**, and **2—5A synthetase** (4). The two last-mentioned enzymes are activated (heavily outlined boxes) by double-stranded (ds) RNA (5). The activated 2–5A synthetase in turn activates the ribonuclease (6), which degrades RNA (7). The activated protein kinase is capable of phosphorylating, and thus inactivating, a factor that initiates synthesis of proteins (8). The end result (9) is inhibition of viral replication.

As well as blocking viral replication, IFNs have profound effects on cells, some of which also help indirectly to control infection. One of the most important is **enhancement of the display on cell surfaces of histocompatibility antigens**, which we have seen are essential to antigen-driven activation of T cells. Others involve modulation of both B and T cell activities, including enhancement of NK and T_c cell cytotoxicity.

In addition to their antiviral effects, IFNs can inhibit cell division and this property, in conjunction with the immunomodulating activities just described, has been used to some effect in treating certain forms of cancer, notably osteosarcoma in young children.

By now, you may be asking two good questions: just how important is interferon in defending against virus infections and is it any use in treating them?

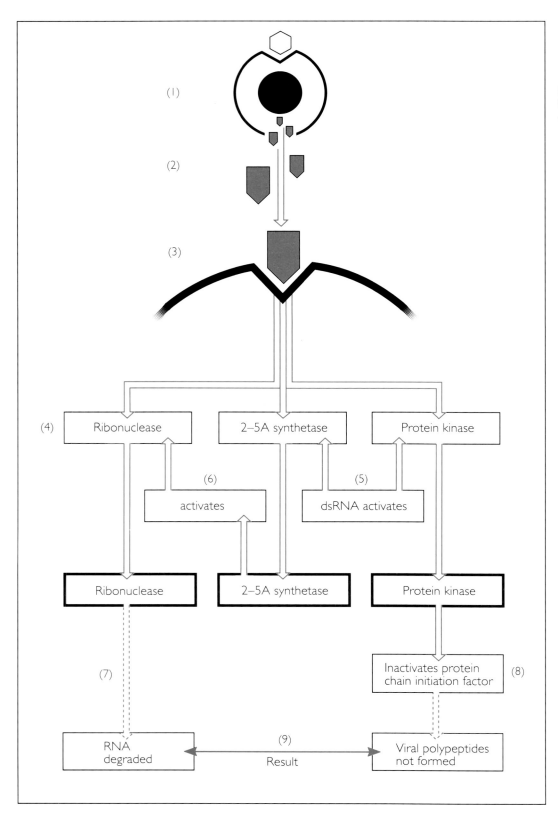

Fig. 5.5. Mode of action of interferon. For explanation see text.

Although the effects of interferon are clear in experiments done in cell cultures, it is quite difficult to disentangle them from all the other components of the immune response in intact animals. IFNs are generated very quickly — within hours of the start of infection — and the consensus of opinion is that they do play a part in the very early stages, holding the fort in the interval before clonal proliferation of T and B cells and antibody production get under way. This view is supported by the finding that treatment of animals with anti-IFN serum exacerbates virus infections; furthermore, some people, who have a natural defect in IFN production, are abnormally susceptible to upper respiratory infections.

With hindsight, the early attempts to use IFN for treatment were doomed to failure, because the amounts available were minuscule by today's standards. Now that genes coding for IFN can be cloned, e.g. in yeast cells, very large quantities can be produced and have been used with fair success in treating chronic hepatitis B; this will be dealt with in Chapter 20, but at this point it is interesting to note that part of the therapeutic effect may be due to IFN-mediated enhancement of MHC display on infected liver cells, resulting in their more efficient destruction by T_c cells. This leads us straight on to the question of immunopathology in virus infections.

3.3 *Harmful immune responses*

Enhanced T cell responses

The destructive potential of the immune system is considerable and can sometimes be dangerous to the host. This was well illustrated during the outbreaks of hepatitis B that used to occur in renal dialysis units before the hazards posed by this infection were properly understood and guarded against. The immunity of the patients was impaired by their illness and by treatment; those who contracted hepatitis B could not therefore provide an effective defence against the infection, which in them tended to develop into the chronic carrier state (Chapter 20). By contrast, members of staff, healthy but presumably exposed to large amounts of virus, mounted vigorous cytotoxic T-cell responses, resulting in autoimmune destruction of their own infected hepatocytes and illness that was, on average, more severe than those suffered by their patients. Similar mechanisms operate in the liver cirrhosis seen in some types of chronic active hepatitis.

Depressed T cell responses

Immune responses to virus infections are complicated by the ability of some viruses to infect and damage the very cells that mediate these reactions. The best example is that of AIDS (Chapter 22), in which the human immunodeficiency virus (HIV) infects and destroys the T-helper subset, which, as we have seen, is crucial in the activation of B cells and of other T lymphocytes; the resultant havoc in the immune system opens the way to so-called opportunistic infections by bacteria and other viruses and sometimes to tumour formation, resulting in death.

Measles is also a good example of the importance of an intact T cell response: in children with a defect in cell-mediated immunity there is no rash, which, as explained in Chapter 4, is the result of cytotoxic T cells destroying virus-infected cells in the skin. The presence of a rash is thus a good sign that the patient is recovering; its absence is sinister, because the likely outcome is uncontrolled viral replication resulting in giant-cell pneumonia and death.

Another example is infection with Epstein–Barr virus, a herpesvirus that causes infectious mononucleosis (Chapter 15) and then becomes permanently latent in B cells. Because of efficient immune surveillance by T cells the latent infection is usually without effect on the health of the host; but an upset of this balance by immune suppression, e.g. in renal transplant patients, can result in transformation of the infected B cells by the virus with uncontrolled proliferation and death from lymphoma.

Viral infection of mononuclear phagocytes (MP)

Many viruses can replicate within these cells, which, as mentioned in Section 3.2, play a major role in regulating immunity. Because of the diversity of viruses involved, of the types and functions of MP, and of the cells with which they interact, the influence of viral infection of MP on the immune response is extraordinarily complex. From the description of the immune system in Section 3.2 you will appreciate that interference with MP by viral infection can, among other activities, affect phagocytosis, antigen presentation, interactions with T cells, chemotaxis, and antibody-dependent cytotoxicity; there follow a few examples of virus–MP interactions that have been elucidated in the laboratory.

The age of the MP may be important: newborn mice are highly susceptible to herpes simplex because the virus multiplies readily in their macrophages; but a pronounced increase in resistance develops about 3 weeks after birth and is related to a decreased ability of the virus to replicate in these cells.

Genetic resistance to certain viruses may be partly mediated by MP.

Influenza and cytomegalovirus infections may depress the function of lung macrophages and increase susceptibility to bacterial infection.

Arginase secreted by macrophages may interfere with the replication of herpes simplex virus, which has a requirement for arginine; this, however, is an unusual type of effect.

The precise role of these and other factors in the immune response cannot as yet be quantified, but they do provide some pointers to the part played in pathogenesis by viral infection of these important cells.

Immune complex disease

In certain chronic infections of animals, immune complexes between virus and antibody lodge in small vessels, particularly in the kidneys, causing glomerulonephritis. This sequence of events does not seem to be an important feature of infections of humans; nevertheless, immune complexes may be the trigger for the disseminated intravascular coagulation seen in some of the viral haemorrhagic fevers (Chapter 19). They may also be implicated in dengue haemorrhagic shock syndrome, which occurs in children with antibody to dengue virus who subsequently become infected with another serotype (Chapter 17).

3.4 *Resistance and recovery*

At the start of this chapter it was emphasized that the mechanisms whereby we resist infection differ from those that are involved in recovery.

By and large, **resistance to invasion by viruses** is mediated by antibodies. These may be:

◆ **passively acquired** by infants from the mother, or by older people given passive immunization, e.g. immunoglobulin to protect against hepatitis A or rabies (Chapter 28);

◆ **actively acquired** as the result of an earlier encounter with the same virus. In this case, the comparatively small amount of existing antibody (**IgG** or **IgA**) is rapidly supplemented by clonal proliferation of antibody-secreting B lymphocytes.

Recovery from infection is a more complicated affair and the responses brought into play depend to some extent on the virus concerned. In the last chapter (Section 3.2) cytopathic viruses were classified as 'bursters', lysing cells with liberation of large numbers of virions into the circulation, or as 'creepers', moving directly from cell to cell. During the viraemia resulting from 'burster' infections such as poliomyelitis or yellow fever, the virions are susceptible to antibody, notably the IgM induced early in a primary infection. Antibody thus plays a larger part in recovery from 'burster' than from 'creeper' infections, the latter being caused by enveloped viruses such as herpes and influenza. 'Creepers' readily induce specific antigens on the surfaces of infected cells, thus evoking powerful cell-mediated responses. Even with lytic virus infections, however, cell-mediated immunity (CMI) is an important factor in recovery.

Experimental evidence for the relative roles of antibody and cell-mediated immunity in resistance and recovery is powerfully supported by observations of infections in patients with deficiencies in these responses. This subject will be dealt with in Chapter 26.

4   REMINDERS

◆ Protection against microbial infection is conferred both by **non-specific** and **specific** (or acquired) mechanisms.

◆ Non-specific factors include **mechanical** and chemical barriers in individuals, and **genetically determined** resistance or susceptibility in populations.

◆ Specific immunity is characterized by: (1) ability of the system to **distinguish between 'self' and 'non-self'** molecules; and (2) **memory**, whereby an enhanced response is evoked to an antigen previously encountered.

◆ The two main components of specific immunity to virus infections are: (1) **antibodies** produced by B lymphocytes in co-operation with antigen-presenting cells and T lymphocytes; and (2) **cell-mediated immunity (CMI)** conferred by a variety of T lymphocytes including particularly helper, suppressor, and cytotoxic cells.

◆ The responses to infections by some viruses are complicated by their ability to **replicate** in **lymphocytes** and **macrophages**.

◆ As a generalization, **protection** against virus infections is largely mediated by antibody, whereas, in **recovery**, CMI is relatively more important, especially in the case of the enveloped 'creeper' viruses.

◆ **Interferons** may help to suppress virus replication during the **early stages of infection** before antibodies and CMI are fully mobilized.

◆ Immune responses can sometimes be **harmful**: for example, a number of manifestations of virus infections are caused by the cytotoxic effects of T cells or by immune complexes of antigen with antibody.

VIRUSES AND CANCER

INTRODUCTION

If any one topic can be regarded as the leading edge in molecular biology, it is the relationship between viruses and malignancy. As with many other aspects of biology, this one was characterized by some important early discoveries, a long lag phase, and then a burst of exponential growth that is still very much in progress. Its ramifications are wide and extend to the fundamental mechanisms by which growth and other activities of cells are controlled at the molecular level. Although the story is now becoming extremely complex, we shall try in this chapter to describe the basic principles as simply as possible.

Table 6.1 lists some of the major advances in the study of tumour formation (oncogenesis), particularly those relating to viral infections. They can be divided into two phases.

1.1 Early researches

This period covers the half century from about 1910 to 1960. It began with the finding by an American scientist, Peyton Rous, that a fowl sarcoma could be transmitted to normal chickens by injecting them with cell-free filtrates of the tumour. Recognition of the importance of this discovery earned him a very belated Nobel Prize. A quarter of a century later, Bittner described a mouse mammary tumour virus (MMTV) that was transmitted to newborn animals in their mothers' milk. This was followed by the identification by Gross and Friend of the two mouse leukaemia viruses that are named after them; it later proved significant (see Section 2.1) that Gross leukaemia had

Table 6.1 Some milestones in the association of viruses with cancer

Early researches

1911	Peyton Rous	Demonstrated that fowl sarcoma is caused by a transmissible agent (Nobel Prize—but not until 1966!)
1927	H.J. Muller	Ionizing radiation mutagenic in *Drosophila*
1936	J.J. Bittner	Discovered mouse mammary tumour virus
1944	O.T. Avery, C.N. MacLeod, and M. McCarty	Showed (in bacteria) that genetic information is contained in DNA
1951	L. Gross	Discovered a 'slow' murine leukaemia virus
1957	C. Friend	Discovered a 'fast' murine leukaemia virus

The modern era

1962	J.D. Watson, F.H.C. Crick, and M.H.F. Wilkins	Worked out the structure of DNA and discovered the genetic code (Nobel Prize)
1963	M. Vogt and R. Dulbecco	Transformed normal cells to tumour cells *in vitro* by exposing them to a virus
1969	G.J. Todaro, R.J. Huebner, and H. Temin	Oncogene hypothesis
1970	D. Baltimore, H. Temin, and S. Mituzani	Discovered reverse transcriptase (Nobel Prize)
1976	D. Stehelin, H.E. Varmus, J.M. Bishop, and P.K. Vogt	Discovered the oncogene Nobel prizes for Varmus and Bishop
1980	B.J. Poiesz, F.W. Ruschetti, M.S. Reitz, V.S. Kalyanaraman, and R.C. Gallo	First isolation of a human leukaemia virus (HTLV–I)

an incubation period of about two years, whereas the Friend agent induced the disease in a matter of weeks.

Many other basically similar agents were identified in a variety of animals and birds, usually causing various types of leukaemia or sarcoma. These agents were referred to as 'oncornaviruses', i.e. *oncogenic RNA viruses*, and classified as type A, B, C, or D according to their morphology and growth characteristics. As with other viruses, infection could be transmitted by passage of infective virions from one host to another. This mode was termed **horizontal** transmission, even when, as with MMTV, it involved transfer from a parent to its offspring, either *in utero* or shortly after birth. Viruses transmitted in this way to previously uninfected cells are termed **exogenous**. But then a new mode was discovered—the transmission of infection from parents to offspring via the germ cells. Part of the viral genome became incorporated in some way into all cells—both somatic and germ-line—of the host, so that it continued to be passed by **vertical** transfer from generation to generation. Viruses transmitted like this are **endogenous**. This type of infection could remain silent, or could result in a high incidence of tumours in a particular strain of laboratory animal. Furthermore, as we shall see, such viruses can, under different conditions, be transmitted either horizontally or vertically. Another important characteristic of these infective agents is that they are not cytolytic.

In addition, it was appreciated early on that papovaviruses (Chapter 21), which contain DNA, could also give rise to tumours, both malignant and benign.

Meanwhile, other discoveries were being made, not directly involving viruses. It was known that ionizing radiation and various chemicals, e.g. small amounts of mustard gas, are oncogenic, and these treatments were also found to induce mutations in fruit flies (*Drosophila*). This led to the idea that cancers might be caused — at least to some extent — by mutations in cellular genetic material, shown in 1944 by Avery and his colleagues to be DNA.

1.2   *The modern era*

This period was heralded in 1953 by the scientific equivalent of a fanfare of trumpets — the discovery by Watson, Crick, and their collaborators of the structure of DNA and the way in which genetic information is encoded within it. Research on the molecular mechanisms of oncogenesis — and on so much else — could now begin in earnest.

Later, Marguerite Vogt and Renato Dulbecco in the USA found that normal cells cultured *in vitro* would, after infection with polyomavirus, take on many of the characteristics of malignant cells (Table 6.2 and Fig. 6.1). This laboratory manipulation, known as **transformation**, was to play a key role in the study of virus-induced cancer.

During the two decades following these researches, more viruses were added to the list of those able to induce malignant changes in cell cultures or in animals. They included adenoviruses, papovaviruses, herpesviruses, and hepatitis B. By contrast, none of the RNA viruses had this property, with the single and obvious exception of the oncornaviruses. The reason became clear with the discovery in 1970 by Baltimore and others of the method by which the oncornaviruses replicate their nucleic acid. This was described in Chapter 2, but, in brief, viral RNA is transcribed into a DNA copy by an enzyme, **reverse transcriptase**, so called because it reverses the usual direction of information transfer from DNA to RNA. As a result of this discovery, the oncornaviruses were redesignated as the family *Retroviridae*, subfamily *Oncovirinae*. (We shall meet other retroviruses when we discuss AIDS in Chapter 22.)

By this time, it was possible to start fitting parts of the jigsaw puzzle together, and, also in 1970, Todaro and others formulated the **oncogene theory**, which later experimental work proved to be fundamentally correct, and which provided the basis for much of our current thinking about oncogenesis.

Table 6.2 Properties of transformed cells

Only one oncogenic virus particle needed to induce a focus of abnormal (**transformed**) cells

Foci are therefore **clonal**, i.e. they originate from a single infected cell

Can be often **propagated indefinitely** *in vitro*

Increased rate of **metabolism** and **multiplication**

Change in karyotype from diploid to **polyploid**

Lack of contact inhibition, with consequent **disorderly growth**

Possess new **tumour (T) surface antigens**

May give rise to **malignant tumours** when injected into animals

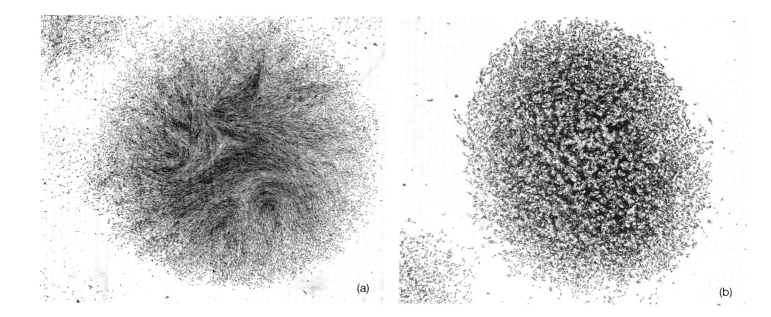

Fig. 6.1. Transformation of a cell line by a DNA virus. **(a)** Normal hamster cells. The colony is flat and the cells are aligned in parallel arrays. **(b)** After infection with polyomavirus the cells are piled up and disorientated. (Reproduced, with permission, from Wyke, J. (1986). In *Introduction to cellular and molecular biology of cancer* (1st edn) (ed. L.M. Franks and N. Teich), p.179. Oxford University Press.)

2 VIRAL ONCOGENES

Knowledge of how retroviruses replicate (see Chapter 2) is essential for following this section. Some, but not all of them, possess a special gene — an **oncogene**, *onc* for short — that is capable of conferring the properties of malignancy on a host cell. Many oncogenes are known, and there are certainly more waiting to be discovered. They work in a variety of ways and there is no point in detailing all of them here. The fundamentals of the process can be illustrated by reference to the first transmissible tumour to be described — the Rous sarcoma (Table 6.1).

2.1 *Mode of action*

Oncogenes are named after the tumours they cause. The one discussed here is *src*, for sarcoma. Figure 6.2 shows (not to scale!) the possible outcomes of infection of a chicken fibroblast with Rous virus (RV). The sequence of events (numbered steps as in Fig. 6.2) is as follows.

1. A retrovirus attaches to specific receptors on the host cell wall.
2. The virion uncoats in the cytoplasm.
3. Viral ssRNA is transcribed to dsDNA by **reverse transcriptase**.
4. Viral DNA (now termed a **provirus**) is integrated into the host-cell genome. The expanded view at the top shows the viral genes.

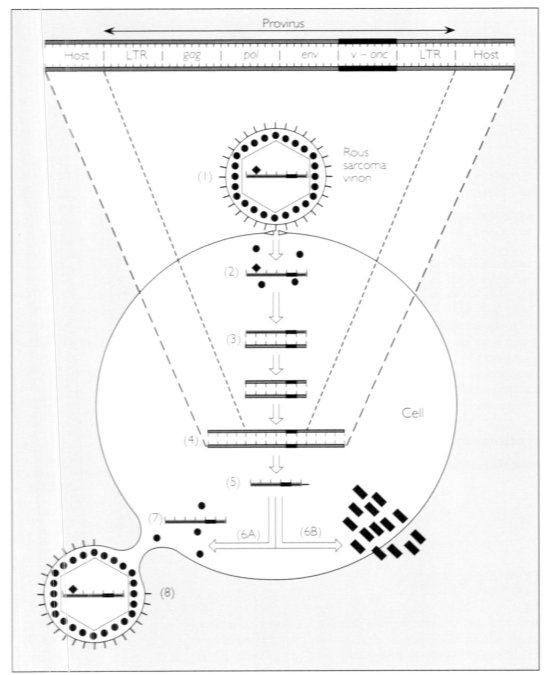

	Capsid protein
	Glycoprotein
	Viral RNA
	Reverse transcriptase
	Viral DNA
	Host DNA
	v - onc
	Protein kinase

Provirus

Host | LTR | gag | pol | env | v – onc | LTR | Host

Rous
sarcoma
virion

(1)

(2)

(3)

(4)

(5)

(7)

(6A)   (6B)

Cell

(8)

Fig. 6.2. Integration of the Rous sarcoma provirus and oncogene into host-cell DNA. For description see text.

The sequence of events from (1) to (4) is similar to that for any retrovirus. At this point, however, one of three courses may be taken.

i.   The integrated provirus often remains quiescent, giving rise to a silent but persistent infection.

ii. It may (5) be translated by a host-cell polymerase to mRNA.

iii. Then *either* the *gag*, *pol*, and *env* genes are transcribed to form new viral components (6a); these are (7) assembled with RNA in the cytoplasm to form (8) progeny virus which is released by budding from the cell surface; *or* the oncogene may become transcribed to an mRNA coding for the oncogene product, in this case a phosphorylating enzyme (protein kinase) that modifies the structure and activity of certain cell proteins. These modifications, in ways not yet fully understood, have as their end result a transformed cell, which, by clonal expansion, forms a malignant tumour. The cells transformed by certain oncogenic viruses (but not Rous sarcoma virus) acquire new tumour (T) antigens in their cell membranes, which may help to change their growth characteristics.

It must be emphasized again that this is just one example of how an oncogene operates; others are integrated in rather different ways, or encode different gene products.

There are pronounced differences in the times taken for the transforming retroviruses to exert their oncogenic effects, The Gross virus (Section 1.1) is an example of a weakly ('slow') transforming retrovirus. These have a full complement of *gag*, *pol*, and *env* genes and thus replicate normally; integration of provirus only infrequently results in transformation. However, some exogenous retroviruses have in the past acquired an oncogene that displaced one or more replicative genes (*pol* in the example shown in Fig. 6.3). Such 'fast' viruses are rapidly oncogenic, but are unable to produce new virions without the aid of a 'helper' virus capable of supplying the missing gene(s).

Host DNA | LTR | *gag* | *onc* | *env* | LTR | Host DNA

Fig. 6.3. Loss of a replicative gene (*pol*) by recombination and consequent displacement with an oncogene (*onc*). For explanation, see text.

3 CELLULAR ONCOGENES

During their research on the *src* oncogene, Stehelin and his colleagues (Table 6.1) made a most remarkable discovery. They found that normal cells of birds, fishes, and mammals contain a counterpart of *src*, having a rather different molecular structure, but coding for the same protein kinase. The same applies to other oncogenes, so that we have to specify whether we are referring to viral oncogenes (v-*onc*), or their cellular counterparts (**proto-oncogenes**, or c-*onc*). It appears that during the course of evolution, viruses have captured (or **transduced**) host genes that in their cellular form participate in the normal activities of the cell, but in their viral form behave as described in Section 2.1. Cellular oncogenes are present in the germ cells of animals, birds, fishes, and even insects.

Can c-onc genes be oncogenic?

Since they occur with such frequency, this question is of crucial importance to our understanding of how cancers in general, not only those related to viral infection, are initiated. The answer is yes; a number of different stimuli may have this effect.

◆ Figure 6.4 shows a provirus, with its *gag*, *pol*, and *env* genes inserted into host-cell DNA. The genes are flanked by long sequences of repeated nucleotides (long terminal repeats, or LTRs), which have regulatory functions. Insertion of the viral LTR may activate an adjacent cellular oncogene (*c-onc*), thus giving rise indirectly to malignant change in the host cell. This mechanism is known as **insertional mutagenesis**.

◆ Carcinogens such as **radiation** and various **chemicals** — including nicotine — may induce oncogenic mutations. These mutations may be very small; experimentally, a *c-onc* gene was made oncogenic by a single base change — a point mutation — in its DNA.

◆ Another way in which cellular genes may be involved is by loss or mutation of a **growth suppressor gene**, normally concerned with controlling cell proliferation.

Host DNA | LTR | gag | pol | env | LTR | c-onc | Host DNA

Fig. 6.4. Activation of a cellular oncogene by insertional mutagenesis. For explanation, see text.

4 ONCOGENIC VIRUSES

It should by now be quite clear why only dsDNA viruses, or retroviruses that synthesize DNA during replication, are oncogenic; only they can perturb host-cell functions in the ways we have described. You will encounter various examples of virus-induced oncogenesis in Parts II and III of this book; Table 6.3 lists the viruses for which there is good evidence of oncogenicity in man. They do not include the following three families of DNA viruses.

◆ The adenoviruses, which do not cause tumours naturally, but only when injected into newborn rodents; they also transform cells in culture.

◆ The parvoviruses, which have no oncogenic potential.

◆ The poxviruses, which give rise only to benign tumours.

Table 6.3 Viruses directly implicated in human cancers

Virus	Tumour	Possible cofactors
Retrovirus		
Human T-cell lymphotropic virus type 1 (HTLV-I)	Adult T-cell leukaemia/ lymphoma	
DNA viruses		
Epstein–Barr virus (EBV)	Burkitt's lymphoma	Malaria
EBV	Nasopharyngeal carcinoma	Nitrosamines in diet; genetic factors
	Immunoblastic lymphoma	Immunodeficiency
Herpes simplex viruses (HSV)	Carcinoma of cervix(?)	
Papillomaviruses	Skin carcinoma	Sunlight
	Carcinoma of cervix	Smoking; other virus infections, e.g. HSV
Hepatitis B (HBV); Hepatitis C (HCV)	Carcinoma of liver	Aflatoxin; alcohol

It is significant that all the viruses listed in Table 6.3 integrate part or all of their genome with that of the host cell. This may result in a latent infection that reactivates with the production of infective virus (Chapter 4), or, if the conditions are right, initiates oncogenic changes. In the case of these viruses, however, the integrated DNA normally codes for viral components, unlike the v-*onc* genes which do not participate in this activity.

5 CANCER – A MULTIFACTORIAL DISEASE

In the sense that many other factors promote tumour formation, oncogenes are not the be-all and end-all of oncogenesis, but in another sense they are. Although we are a long way from the definitive answers, it seems increasingly likely that they represent a final common pathway through which these other factors eventually act. There is more than one stage in oncogenesis. The establishment of a cancer requires first an **initiation** event. This is an irreversible change in a cell, which does necessarily result in tumour formation: insertion of an oncogene or other viral DNA is an example. The next step is the action of a **promotor**, i.e. some influence that pushes the cells toward the malignant state. Some or all of the possible cofactors listed in Table 6.3 may act in this way.

Host factors

In thinking of oncogenesis at the cellular level, we must not forget the influences exerted by the responses of the host as a whole. Thus, **immune mechanisms** play an essential part in determining the course of malignant disease. It is noteworthy that cancers are more liable to occur in immunodeficient patients (Chapter 26), perhaps because the immune surveillance system loses its capacity to recognize cells with new tumour antigens as 'foreign'.

Some forms of immunosuppression are themselves due to viral infection, both by retroviruses and others. In this way, viruses can promote oncogenesis by **indirect** means. The human immunodeficiency viruses provide an example (Chapter 22). They do not themselves possess oncogenes but, by inducing profound immunodeficiency, permit the growth of tumours that would otherwise be suppressed. Immunosuppression by heavy malarial infection may account for the high prevalence of Burkitt's lymphoma (Fig. 4.1(c)) in large areas of Africa (Chapter 15).

Genetic influences determine susceptibility to some forms of cancer; for example, nasopharyngeal carcinoma following Epstein–Barr virus infection is particularly liable to occur in people of southern Chinese origin. Dietary and hormonal factors may also be implicated in some malignancies (Table 6.3), but their modes of action are not fully understood.

6 REMINDERS

◆ Some viruses that possess a **DNA genome**, or that **synthesize DNA during replication** (the retroviruses), are able to induce malignant changes in cells, both *in vivo* and *in vitro* (transformation). RNA viruses other than retroviruses are not oncogenic.

◆ Transformed cells have an **increased rate of multiplication**, grow in a **disorderly fashion**, and can be **propagated indefinitely** in the laboratory; some possess **tumour (T) antigens** on their cell surfaces.

◆ Malignant changes can be initiated by a variety of **mutagenic stimuli**, including radiation, chemical carcinogens, and insertion of viral DNA into the host-cell genome.

◆ In the case of **retroviruses**, the DNA is a copy of the RNA genome — the **provirus**. Oncogenic retroviruses possess an additional gene — the **oncogene** — which may either remain quiescent or be translated to proteins that affect the normal growth characteristics of the host cell; it does not participate in synthesis of viral components.

◆ Viral oncogenes (v-*onc*) possess almost identical **cellular counterparts** (c-*onc*) from which they were derived in the distant past. These c-*onc* genes can also be activated by mutagenic stimuli to induce malignancy. They may represent a final common pathway through which various oncogenic factors operate.

◆   Some DNA viruses are oncogenic. In their case the integrated DNA is not a specialized oncogene, but encodes viral components in the normal way.

◆   Viral oncogenesis may be assisted by **host factors**, including immuno-suppression, genetic make-up, dietary habits, and exposure to mutagenic influences such as ionizing radiation or chemical carcinogens.

CHAPTER SEVEN

VIRUSES AND THE COMMUNITY

1 INTRODUCTION

In the first two chapters, we considered viruses at the molecular level, after which we went on to discuss them in relation to individual cells, to organs of the body, and thence to their effects on individuals. It is now time to stand back and look at a much broader canvas — the way in which viruses affect whole communities. This is the province of **epidemiology**, a word of Greek derivation meaning 'upon (i.e. affecting) the people'; the corresponding term for disease in animal communities is **epizootology**.

The story of the way in which microbes and other parasites affect communities is one of a constantly shifting balance of power between parasite and host, both of which, during their evolution, evolve quite elaborate means of attack and defence and thus of survival. The results of such battles — which determine the effects of microbes on individuals and hence on the community — are the outcomes of a highly complex interplay of factors. These were neatly summarized by the American epidemiologist, John Paul, under the headings of *seed*, *soil*, and *climate*, the first referring to the parasite, the second to the host in which it grows, and the third not just to the weather, but to all the other environmental factors involved.

2 DEFINITIONS

Prevalence The number of cases of a given disease—clinical or subclinical—recorded **at a particular time** and expressed as a proportion of the population under study.

Incidence The number of cases recorded **during a particular period**, e.g. one year. This measurement is often given in terms of an **attack rate**, i.e. the number of cases per thousand or per hundred thousand of the general population (or of a subgroup within it) during the period in question.

Endemic This term refers to an infection that is constantly present at a significant level within a community. Herpes simplex is an example. Endemicity may be high or low. The corresponding term for infections of animals is **enzootic**.

Epidemic An unusual increase in the number of cases within a community. In this context, 'unusual' is defined arbitrarily: 100 cases of measles in London might not be regarded as an epidemic, but 100 cases of Lassa fever certainly would be. Localized epidemics are usually referred to as **outbreaks**. The corresponding term for infections of animals is **epizootic**.

Pandemic An epidemic involving several continents at the same time. The corresponding term for infections of animals is **panzootic**.

3 WHAT USE IS EPIDEMIOLOGY?

There is much more to this science than dry tables of statistics; it is the key to four major clinical activities, the first three of which may be aided by the application of mathematical models.

◆ **Predictions of trends in diseases.** For example, knowledge of the behaviour of an infection in the past helps to predict the course of an epidemic.

◆ Guide to the **introduction, improvement, or modification of control measures**, e.g. immunization programmes, control of insect vectors, or improvements in hygiene.

◆ **Evaluation of the success of such measures**, locally, nationally, or world-wide.

◆ **Aids to diagnosis.** A knowledge of what infectious diseases are currently prevalent is very useful to the physician as a diagnostic pointer; it may also suggest to the laboratory the need for certain specific tests.

4 EPIDEMIOLOGICAL METHODS

It is not possible here to discuss epidemiological techniques in detail. For our purposes, it is enough to know that, for infective diseases, the two principal methods of surveillance are **clinical** and **microbiological**, both of which must be tied in with an adequate system of **data collection** and **processing**.

4.1 *Clinical observations*

Well before bacteria and viruses were discovered, clinical observations alone, carefully recorded, made major contributions to epidemiology. The story is well known of how, many years before the cholera vibrio had been identified, Dr John Snow determined which water companies supplied different parts of London, and concluded that this infection was waterborne. In 1854, he identified the Broad Street pump in Soho as the source of a major outbreak of cholera and dealt with the situation very successfully by persuading the parish authority to remove the pump handle! There are many examples of equally acute observations of viral infections, one of the most striking being P.L. Panum's study of measles in the Faroe Islands in 1846. During that year there was a major epidemic in this isolated community, which, not having been exposed to the infection since 1781, consisted almost entirely of non-immune people. Within the next 6 months, more than 6000 people in a total population of 7782 contracted measles, an extraordinary attack rate of over 770 per 1000. Panum noted that none of the elderly people who had had measles in the previous epidemic acquired it on this occasion. By meticulously recording the dates of contacts and of the onset of disease, this young Danish doctor established that measles is infective for others only at about the time of appearance of the rash; he found the incubation period to be 13–14 days and confirmed by personal investigation that cases with significantly longer or shorter incubation periods either were suffering from something other than measles or had had a contact outside this period.

Nearer our own time, N.M. Gregg noticed that during 1941 there was a high incidence of cataract in newborn babies in Sydney, and that this abnormality was often associated with deformities of the heart. He then made the acute observation that the mothers of most of these infants had had rubella (german measles) while pregnant during an epidemic in the previous year. The making of this association, purely by clinical observation, was fundamental to the recognition that several virus infections acquired during pregnancy may damage the fetus. We shall return to this subject in Chapter 25.

4.2 *Laboratory studies*

Good clinical observations are useful, provided that an adequate **case definition** is established at the outset; this means laying down the criteria by which a case is accepted or rejected as suffering from a particular disease. Purely clinical studies do, however, have the following disadvantages.

◆   Despite the use of case definitions, some syndromes, such as respiratory infections or diarrhoea, are often difficult to diagnose accurately on clinical grounds alone.

◆   They cannot reveal the extent of very mild or subclinical infections.

◆   They usually cannot provide accurate information about the prevalence of a particular viral infection in the past (which is often a good guide to the state of immunity to that virus possessed by the community as a whole).

For these purposes, laboratory investigations are brought into play, and are especially useful in viral infections, most of which induce a good and long-lasting antibody response. As in general diagnostic work, the tests fall into two main categories:

◆   detection of virus in acutely ill patients;
◆   detection of specific antibody.

For epidemiological work, the first category is of limited value; it is useful for some purposes, e.g. identifying the prevalent strain of influenza during an epidemic, but such tests are normally done only on a limited sample of the population. Serological tests, many of which are now automated or semi-automated, are much more practical for large-scale surveys.

5 SEROLOGICAL EPIDEMIOLOGY

This expression was also coined by John Paul, who wrote that serum antibodies 'represent footprints, either faint or distinct, of an infection experienced in the remote or recent past'. During the 1930s it was found, even with the crude methods then available, that antibodies to certain infections were present in much higher proportions of the study populations than was warranted by the amount of clinically apparent disease. This situation can be likened to that of an iceberg (Fig. 7.1) in which the part above water represents clinical illness, and that below, the prevalence of antibody in the population. The simultaneous discovery that the prevalence of antibody varies with age provided a powerful tool for studying the history of viral infections in communities.

Now look carefully at Fig. 7.2, which contains much more information than at first appears. These are the results of early studies on antibody to a poliovirus, carried out during 1949–51 in three locations differing greatly in socio-economic conditions. The first, in Cairo, included young infants, and clearly shows that those aged less than 6 months still had maternal antibody, which was lost after this period; thereafter, however, it rapidly reappeared, reflecting an active immune response to early infection with the virus. This

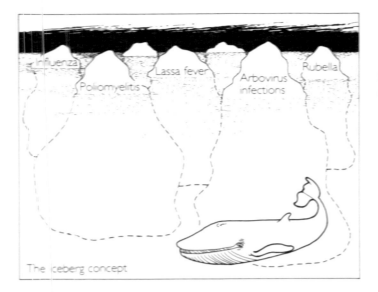

Fig. 7.1. The iceberg concept of infections in communities.

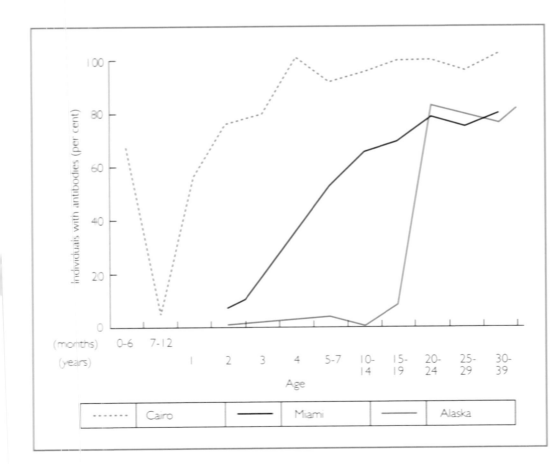

Fig. 7.2. Prevalence of poliomyelitis antibody in different populations. Percentages of individuals with antibodies to poliovirus type 2, surveyed during 1949–51. (Reproduced with permission from Paul, J.R. and White, C. (ed.) (1973). In Serological epidemiology, p.10. Academic Press, New York.)

finding was surprising, because epidemics of *clinical* poliomyelitis were unknown—or at least unreported—in Egypt; it represented hitherto unsuspected infection—the part of the iceberg under water. The early acquisition of antibody is characteristic of societies with poor hygiene in which transmission of infections by the faecal-oral route readily occurs.

In Miami, a world apart both geographically and economically, hygienic conditions were such that exposure to the virus came later in life, and poliomyelitis tended to occur in clinically apparent outbreaks at intervals of 7–10 years. (We explain in Chapter 14 how at certain times and places poliomyelitis is more severe in older than in younger people.)

In the much more isolated Alaskan population, there had been an epidemic of poliomyelitis 20 years previously, but none since. Note that the percentage of people with antibody is insignificant in those aged less than 20 years, but rises steeply to nearly 80 per cent in those aged 20 years or more. If Dr Panum had had the benefit of serological methods, he would have found a very similar pattern 20 years after the measles epidemic in the Faroes in 1846.

5.1 *Methods*

In carrying out serological surveys it is important to **define the objectives**, to use a **suitable sample** of the study population, and to gather **adequate information** about those tested. If these conditions are met, it is often possible to use the sera for multiple studies, not only on different infections, but also, for example, on blood chemistry. Sera stored at $-20°C$ or below can be kept for many years and used for retrospective investigations. Drops of blood can be collected on discs or strips of filter paper, dried, and later eluted in buffer for antibody tests; this technique is especially useful for young infants.

The methods of measuring antibodies vary with the virus concerned; for example, those to polioviruses, measles, and influenza might be assayed, respectively, by neutralization, complement fixation, and haemagglutination-inhibition tests (see Chapter 27).

5.2 *Monitoring an immunization programme*

In addition to studying patterns of disease, serological tests are invaluable for monitoring the success of mass immunization programmes. Here, it is important to know what proportion of people in the target population is immune to the disease in question, whether as a result of the immunization itself or of natural infection. The following is an example of what can be done at a purely local level.

In the UK, sera from women attending antenatal clinics are routinely screened for rubella antibody; if negative, the patient is advised that she should be immunized soon after delivery in order to avoid the risk of rubella in subsequent pregnancies with consequent damage to the fetus (see

Chapter 25). During 1983, it was noticed in two London virology laboratories that the prevalence of rubella antibody in women from the Indian subcontinent was significantly lower than that in the indigenous population. This finding indicated the need for a stepped-up information and immunization campaign directed at this particular group.

5.3 *Processing of data*

The collection of epidemiological information is pointless unless there is an efficient system for collating it and applying the results to improvement of the public health. Such systems exist at the local, national, and international levels. It is not possible here to explain in detail the machinery for processing epidemiological information, but an important principle is that much of the traffic is two-way; data collected by clinicians and in laboratories are processed centrally, and the results are ultimately returned to these 'front-line' workers for action at the local level.

On the international level, much information is collected by the World Health Organization, an outstanding example being the monitoring, through national centres, of the incidence of influenza and the prevalent strains of virus, an activity of prime importance to vaccine manufacturers.

In the USA, the surveillance and investigation of infective diseases is managed centrally by the Centers for Disease Control (CDC) in Atlanta, Georgia. In the UK, these functions are the responsibility of the Public Health Laboratory Service (PHLS), which maintains the Communicable Disease Surveillance Centre in London and a network of microbiological laboratories covering the UK. In addition, clinical data are collected by panels of physicians who report on specified diseases through the Royal College of General Practitioners. There are also weekly reports on certain diseases to the Office of Population Censuses and Surveys; and certain infections must by law be reported to the appropriate authorities (Appendix B). Other countries have analogous systems, their scale and quality being largely governed by economic circumstances.

In some areas of the USA, there are the admirable Virus Watch Programs, in which volunteer families are observed and systematically tested over periods of years for evidence of enteric and respiratory virus infections.

6 FACTORS IN THE SPREAD OF VIRAL INFECTIONS

At this point, we must take stock of the 'seed', 'soil', and 'climate' factors mentioned in Section 1. First, the seed.

6.1 *Characteristics of the virus*

Table 7.1 lists the main features of viruses that determine how they are transmitted and hence their potential for spread within communities.

First, how well does the virus survive in the environment on its way from one host to another? Enteroviruses such as hepatitis A and polioviruses can remain viable for weeks in water or sewage, an obvious advantage for waterborne agents. They are also unusually resistant to acid pH, which helps them to survive transit through the stomach on their way to their site of replication in the small intestine. By contrast, the rhinoviruses — also members

Table 7.1 Properties of viruses that determine transmissibility

Property	Particular features	Examples
Survivability outside the host	Resistance to ambient temperatures, drying, ultraviolet light (in sunlight), pH	Enteroviruses, e.g. polio and Coxsackie viruses
Existence of an alternative host	If so, direct transmission? via arthropod vector?	Rabies virus Arboviruses, e.g. yellow fever
Portal of entry	(see Tables 4.1 and 4.2)	
Evasiveness	Rapid multiplication on a mucous membrane before immune response can be mounted	Viruses infecting the respiratory tract, e.g. rhinoviruses, and conjunctiva e.g. some adenoviruses
	Variability in antigenic structure, thus evading immune response to a previous infection	Influenza A and B viruses, human immunodeficiency virus (HIV)
Pathogenesis	Incubation period: short, medium, or long	See Chapter 4
Route by which virus is shed	Respiratory secretions	Viruses causing childhood fevers, e.g. measles, mumps rubella; those causing respiratory infections
	Conjunctival secretions	Conjunctivitis viruses, e.g. some adenoviruses, enterovirus 70
	Skin, epithelial mucosae	Warts, herpes simplex and zoster
	Faeces	Entero- and rotaviruses

	Blood
	  Transfusion
	  contaminated needles or instruments
		Cytomegalovirus, hepatitis B and C, HIV

of the *Picornaviridae* — being spread by the respiratory route, do not possess this property.

The existence of a **reservoir of infection** in a primary (alternate) host, usually a mammal but sometimes a bird, clearly has many implications for the mode of spread. Such infections are called **zoonoses**, among which are many infections by toga- and flaviviruses (Chapter 17) and by those causing certain haemorrhagic fevers (Chapter 19). If — as is often the case — a vector such as a mosquito or tick is also involved, many more factors complicate the picture; these include its feeding and breeding patterns, range of mobility, length of time for which the virus persists within it, and whether the virus is transmitted to its offspring. The transmission of such infections to humans clearly depends on the degree of exposure of the latter to the vector, which in turn may be conditioned by the place of work or recreation.

The routes by which viruses enter and are shed from the body are obviously important in transmission, and what happens to them within the host may be equally so. Much depends on how well the virus is able to evade the host's defences, in terms either of its site of replication or of its ability to undergo mutations that give rise to new, antibody-resistant strains. The ways in which viruses spread within their hosts were discussed in Chapter 5; they determine the incubation period of a given infection and thus the rapidity with which it can be transmitted from person to person. To take two extremes, conjunctivitis caused by enterovirus type 70 has an incubation period of about 2 days and causes explosive epidemics that sweep through whole communities like wildfire; the AIDS viruses, with incubation periods measured in years, spread correspondingly much more slowly and insidiously.

6.2 *Characteristics of the host and the environment*

We have combined the 'soil' and 'climate' factors in Table 7.2 because one cannot consider host species apart from their environments. We have not given examples of how the various factors operate because there are so many that isolated instances would give a misleading impression. There is, however, one very important point. The environment does to a great extent determine what sort of virus infections are most prevalent in given geographical areas; this means that there is a considerable difference between the patterns of infection in developed and developing countries (Table 7.3).

Table 7.2 Characteristics of the host and environment that influence the pattern of viral infections

The host

Age
Sex
Ethnic group and genetic factors
Occupation and economic status
Nutrition
State of immunity

The environment

Geographical location
Urban or rural setting
Existence of zoonotic infections/vectors
Socio-economic status/state of hygiene/overcrowding

Table 7.3 High-prevalence viral infections: comparison between developed and developing countries

Developed countries

Upper respiratory tract infections
Paramyxoviruses
Influenza
Herpes viruses
Papillomaviruses
Human immunodeficiency virus

Developing countries

All the above, plus

Poliomyelitis
Gastroenteritis viruses
Hepatitis A
Hepatitis B
Yellow fever; haemorrhagic fevers and encephalitis due to arbo-, filo-, and arenaviruses
Rabies

7 HERD IMMUNITY

This expression signifies the proportion of the population that is immune to a given infection, whether as a result of natural infection or artificial immunization. Most acute viral infections induce firm and long-lasting immunity;

infective but apparently healthy carriers are not usually a significant factor in the behaviour of endemic or epidemic viral infections.

There are two main methods by which viral infections are propagated in communities.

1. **From person-to-person.** Viruses spread in this way are usually transmitted by the respiratory route, e.g. influenza, colds, measles, rubella, mumps.
2. **From an external source.** Many viral infections are spread by the faecal–oral route (e.g. poliomyelitis, hepatitis A, gastroenteritis viruses) or by the bites of infected arthropods (e.g. tickborne encephalitis, yellow fever, dengue fever).

In both modes of spread, the degree of herd immunity is of paramount importance in determining patterns of endemicity and epidemicity. From what has been said, it will readily be appreciated that such patterns are governed by extremely complex factors, so much so that they are often difficult to predict even with the most elaborate mathematical models. Nevertheless, three simplified examples will give a good idea of the general principles.

7.1   *An epidemic in an isolated community*

First, let us go back to the 1846 measles epidemic in the Faroe Islands. The situation faced by Dr Panum is shown schematically in Fig. 7.3. The box represents the isolated community, all of whom, except for a small minority

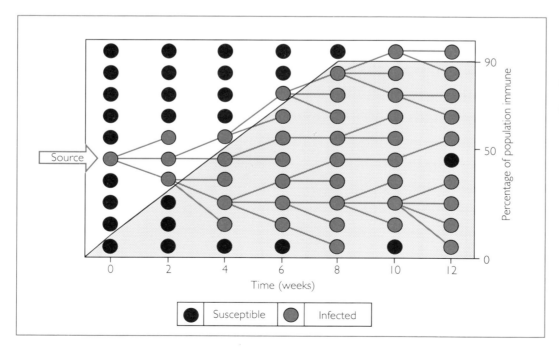

Fig. 7.3. Spread of measles in a highly susceptible population in the Faroe Islands, 1846.

of elderly people, had never been in contact with this infection and were thus fully susceptible. The arrow at the left represents the cabinet-maker who imported the infection on returning from a trip to the mainland. The incubation period was 2 weeks and, on the assumption that those who became infected were immune after another 2 weeks, the increase in herd immunity can be plotted, and is shown as the shaded area. (Needless to say, epidemics do not behave as regularly as this in real life. Successive waves of infection soon get out of synchrony.) Note that some infections are 'dead-end', i.e. are not transmitted to anyone else. The epidemic petered out when there were too few susceptible people left to keep the chain reaction going. In practice, not all the population must be immune for this to occur; it can happen when the proportion reaches 70–80 per cent.

7.2   *Endemic infections*

Look again at Fig. 7.2, in which the age-specific distributions of antibodies to poliovirus type 2 in different geographical areas during 1949–51 are compared. In Cairo most of the population had been infected during infancy, an age when most polio infections are asymptomatic. Since immunity following exposure to this virus is virtually life-long, it follows that the proportion of susceptibles in the total population was very small (Fig. 7.4(a)). There may have been the odd sporadic case of clinically apparent disease, but the pie chart shows that there were certainly not enough susceptible people to sustain

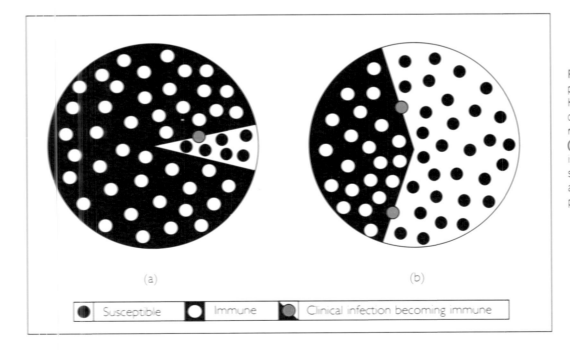

(a)   (b)

| ● Susceptible | ○ Immune | ◑ Clinical infection becoming immune |

Fig. 7.4. Herd immunity in poliomyelitis. **(a)** 'Cairo' model. Herd immunity is high and clinically apparent infection is rare: no danger of epidemic. **(b)** 'Miami' model. Herd immunity is only moderate and there are sporadic cases of clinically apparent infection: an epidemic is probable.

an epidemic. By contrast, the proportion of susceptible people in the various age groups in Miami did not fall below 50 per cent until late adolescence, a period of life when poliomyelitis is liable to cause severe paralytic illness. The situation in this age group is shown in Fig. 7.4(b), from which it can be seen that there has been a build-up of susceptibles to a point at which an epidemic could be initiated. Before the introduction of polio vaccine, such epidemics did, as mentioned in Section 5, occur at 7–10 year intervals in this community.

7.3   *The periodicity of epidemics*

In the absence of interference with the natural order, e.g. by introducing vaccines, improvements in hygiene, or campaigns against arthropod vectors, the pattern of acute infections may alternate between endemic and epidemic. Epidemics create highly immune populations, which, however, steadily become diluted with susceptibles as new babies are born into the community. When the proportion of non-immune people reaches a 'critical mass', contact with an infectious person may start a 'chain-reaction' epidemic; water- or vectorborne diseases may also occur in epidemic form as a result of waning immunity in the general population. In areas unaffected by immunization, epidemics of acute infections such as measles or rubella may recur at regular intervals. Figure 7.5 shows such a pattern and also illustrates the effect of introducing immunization. Another type of periodicity, superficially similar but basically different, is shown in Fig. 7.6. Here, regular annual epidemics result from infections transmitted from an external source — in this case food — and not from person to person. The peaks are due not so much to a build-up of susceptibles, but to the increased opportunities for bacteria to multiply and be transmitted during the warmer months.

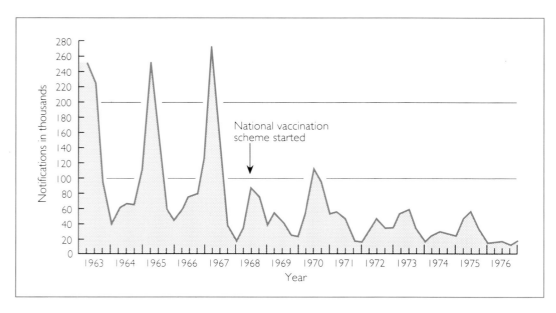

Fig. 7.5. The periodicity of measles epidemics. This pattern typically occurs when the infection is transmitted from person-to-person and confers long-lasting immunity. The effect of introducing mass immunization is apparent. (Reproduced by permission of the Public Health Laboratory Service Communicable Disease Surveillance Centre.)

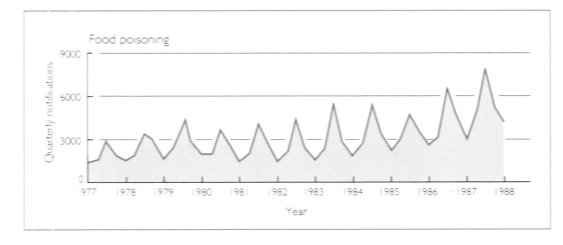

Fig. 7.6. The periodicity of food-poisoning epidemics. This pattern results from periodic outbreaks of 'common-source' infections. (Data from Figure A, OPCS Monitor, MB2 89/1 of 18 April 1989. Reproduced by permission of the Office of Population Censuses and Surveys.)

8 REMINDERS

◆ Epidemiological techniques are essential for **predicting health trends** in the community and as a guide to the **implementation and evaluation of control measures**.

◆ Careful clinical observations, making use of adequate **case definitions**, are important in establishing patterns of endemic and epidemic disease, but usually need supplementing by laboratory tests.

◆ **Serological surveys** of communities must be well designed in terms of their objectives and collection of adequate information about the study population. They show that for many viral infections there are high ratios of antibody positivity — the 'footprints' of past infections — to clinically apparent disease.

◆ To be of value, **epidemiological information** must be properly **collected** and **collated** at local, national, and international levels.

◆ Many factors affect the **transmission** of viral infections, among which are **survivability** outside the host, the existence of **reservoirs of infection** in alternate host species, the need for an **arthropod** vector, the routes by which the virus **enters** and **is shed** from the host, and **incubation period**.

◆ The prevalence of many viral infections differs with **socio-economic status**, and between developing and developed countries.

◆ The degree of **herd immunity** is important in determining the balance between endemicity and epidemic disease. In turn, shifts in this balance may result in recurrences of epidemics of a particular disease at regular intervals.

SPECIFIC INFECTIONS

2.1 *Properties of the viruses*

Classification

Members of the *Adenoviridae* family are icosahedral DNA viruses some 80 nm in diameter. Both mammalian and avian adenoviruses are known and form two genera. The only reason to remember some of the avian adnoviruses is that they may inadvertently be present in eggs used for virus vaccine production and at least one, CELO (*c*hick *e*mbryo *l*ethal *o*rphan virus), causes cancer in animals. Forty-one human adenoviruses are now known, but this number will undoubtedly creep slowly upwards as time passes. They are divided into six subgenera (A–F), mainly on the basis of the nucleotide sequences of the DNA. The degree of gene homology between members of a particular group ranges from 50 to 100 per cent and the viruses falling into a given subgenus tend to have similar pathogenic and epidemiological properties (Table 8.1).

Table 8.1 Classification of human adenoviruses and associated clinical symptoms

Subgroup	Representative viruses	Target organ	Epidemiology
A	12, 18, 31	Gastrointestinal tract	Endemic
B	3, 7, 11, 21	Pharynx, lungs, urinary tract, conjunctiva	Epidemic
C	1, 2, 5, 6	Pharynx	Latent throat infection
D	8, 9, 19	Eye (keratoconjunctivitis)	Epidemic
E	4	Upper respiratory tract	Epidemic
F	40, 41	Gastrointestinal tract	Endemic

Morphology

When the first electron microscope pictures of adenoviruses were published by Horne and his team at the National Institute for Medical Research in London, they caused a minor sensation, revealing as they did an architecture of remarkable beauty and precision (Figs 1.6 and 8.1). The capsid is formed from 252 capsomeres, which are arranged in a icosahedron with 20 sides and 12 vertices. A unique morphological feature is a slender fibre projecting from each of the 12 vertices of the icosahedron, giving the virus something of the appearance of an orbiting satellite.

Replication

After entry by the 'sliding' mechanism and uncoating, the core is transported to the nucleus, where replication, which in general follows the pattern for a dsDNA virus (Chapter 2), takes place.

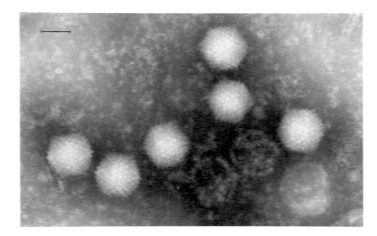

Fig. 8.1. Electron micrograph of an adenovirus (by courtesy of Dr David Hockley). Scale bar, 50 nm.

Transcription of both strands of the viral DNA leads to sequential formation of early and late mRNAs, coding respectively for 'early' and 'late' proteins. About 12 non-structural early proteins are formed. The structural proteins are synthesized late, following DNA replication.

New virus particles are assembled in the nucleus, often in such vast numbers that they form crystalline aggregates. Not unexpectedly, with this amount of viral synthesis the synthetic machinery of the cell is progressively shut down, especially during the latter part of the viral cycle, and the cell dies.

2.2   Clinical and pathological aspects

Endemic respiratory infections in children

Most children are infected with adenoviruses early in life, but probably fewer than half these infections result in disease, the frequency of symptomatic illness depending on the type of virus; adenovirus 2, for example, causes comparatively few. In young children the symptoms include a stuffy nose and cough, whereas in older children pharyngitis is common.

Epidemic disease in military recruits

Adenoviruses are notorious for causing outbreaks of upper and lower respiratory tract infections in military recruits, probably because of crowding and stress. The illness usually lasts for about 10 days and pneumonitis is not uncommon.

Pharyngoconjunctival fever

This syndrome is characterized by pharyngitis and conjunctivitis, mainly in children and young adults. Many outbreaks have been associated with swimming pools, and the possibility of an adenovirus infection should be considered in any patient with conjunctivitis developing about a week after using a communal pool.

Epidemic keratoconjunctivitis ('shipyard eye')

Why 'shipyard eye'? This form of keratoconjunctivitis, which may leave permanent corneal scars and impaired sight, first occurred in epidemic form during the Second World War in American shipyards, which were, of course, very busy. Foreign bodies, e.g. flakes of rust, often had to be extracted from the workers' eyes and the infection was spread both within and without the clinics that they attended. In one shipyard, a particular employee acquired a high reputation for removing foreign bodies more skilfully than the local doctors; unfortunately, he never sterilized his home-made instruments and thus spread adenovirus type 8 to many hundreds of his workmates. Any criticism of this amateur operator should be tempered by the knowledge that, on a number of occasions, adenovirus infections of the eye have been transmitted by medical staff using infected solutions or inadequately sterilized instruments, notably tonometers.

Gastrointestinal infections

Symptomatic adenovirus infections of the gut occur mainly in infants and are caused by types 40 and 41. These agents do not grow in cell cultures (see Chapter 11).

Other syndromes

Adenoviruses have been implicated in acute intussusception in infants, necrotizing enterocolitis, acute cystitis, and meningoencephalitis. They may also cause life-threatening pneumonia and other infections in immunocompromised patients, including those with AIDS.

Pathogenesis

It was rather difficult to establish that adenoviruses actually cause respiratory illness, since they are found in many apparently normal tonsils and adenoids and may be shed by children over long periods without causing disease.

During a relatively short incubation period of 5–10 days the virus may replicate in the pharynx, conjunctiva, and small intestine. In immunosuppressed patients generalized infections can occur and the virus (often adenovirus type 7) is found in the lungs, brain, and kidneys.

Epidemiology

It is now recognized that all 41 adenovirus serotypes are endemic in the community and that some may, for unknown reasons, cause explosive outbreaks of disease, usually respiratory, but also involving the eye. Another feature of the epidemiology of adenoviruses is the degree of seasonal variation. As an example, most outbreaks of pharyngoconjunctival fever in school-age children occur in the summer, perhaps because they use swimming pools more often. Eye infections may also be acquired following infection of the respiratory or alimentary tracts and, as we have seen, by inadequately steril-ized instruments or other fomites. Epidemics of respiratory disease in military recruits occur almost exclusively in the winter. In general, of the 41 adenovirus serotypes, infections with types 2, 3, 5, and 7 are most common throughout the world. Type 1 and 2 infections occur in early childhood, whereas types 3 and 5 predominate later in life. The virus is known to infect persons by aerosol and by eye contact, but is probably also spread by the faecal–oral route, especially where hygiene is poor.

Laboratory diagnosis

Few laboratories carry out serological investigations with adenoviruses; thus laboratory help depends on virus isolation or on rapid identification of virus-infected cells in clinical samples by immunofluorescence.

Viral isolation from faeces, pharyngeal swabs, conjunctival swabs, or urine is slow, taking at least a week and perhaps as long as a month with some serotypes, e.g. those of group D.

Human diploid, HeLa, or Hep-2 cells can be used for viral isolation and observed for typical CPE. The pH of the medium usually falls rapidly as the virus-infected cells become swollen, rounded, and refractile, clustering together like bunches of grapes. This is particularly noticeable with adenoviruses of subgroups B, C, and E.

Once a virus has been isolated it may be subgrouped by agglutination patterns with rat or rhesus monkey erythrocytes or by haemagglutination-inhibition (HI) tests with specific antisera.

Prophylaxis

Because of the disruptive effects of large outbreaks of respiratory disease in army camps, the US military as long ago as the 1960s encouraged the development of a vaccine. Live preparations of adenovirus types 4 and 7 were enclosed in a gelatin capsule and swallowed, bypassing the stomach and being released in the intestine. Here the virus replicates and induces immunity, but causes no overt disease. But the problems for widespread use of such a vaccine in the community are formidable, not least of which is the variety of serotypes causing respiratory disease. Another worrying thought is the oncogenic effects of certain adenoviruses in animals, although there is no evidence of such effects in humans.

3 CORONAVIRUSES

Coronaviruses (Latin *corona*: a crown) infect humans, animals, and birds. The two known human strains infect the respiratory tract and are confined, normally, to the ciliary epithelium of the trachea, nasal mucosa, and alveolar cells of the lungs. Some of the first isolates were obtained in the UK from volunteers at the MRC Common Cold Research Unit, Salisbury.

3.1   *Properties of the viruses*

Classification

Within the family *Coronaviridae*, two serotypes of human coronavirus are known; the remaining 10 identified viruses infect animals and birds. All the viruses are pleomorphic, and have a lipid envelope through which spikes penetrate to give a characteristic appearance. The genome is ssRNA with positive polarity.

Morphology

Coronaviruses are 60 to 220 nm in diameter and have club-shaped surface spikes about 20 nm in length. These spikes give the virions the appearance of a crown (Fig. 8.2).

In thin sections the envelope appears as inner and outer shells separated by a translucent space. Coronaviruses contain two major envelope proteins. The first, the matrix protein, is a transmembrane glycoprotein. The second spike protein, which constitutes the surface peplomer, is responsible for eliciting neutralizing antibodies. The internal ribonucleoprotein (RNP) component has the appearance of a helix condensed into coiled structures of varying diameter.

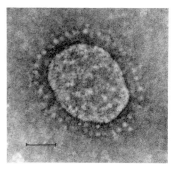

Fig. 8.2. Electron micrograph of coronavirus. The surface spikes protrude through the lipid bilayer to give the appearance of a corona, or crown. (By courtesy of Dr Ian Chrystie.) Scale bar, 50 nm.

Genome and polypeptides

The genome is positive-sense ssRNA and is 15 000–20 000 nucleotides in length. Virions initially attach over the whole cell surface, but are then rapidly redistributed away from the cell periphery by an energy-requiring process. Uptake into cells is rapid and temperature-dependent, probably involving viropexis. The essential feature of coronavirus genomic replication—which is unique for an RNA virus—is mediation of genetic information through multiple subgenomic mRNAs, each of which directs the translation of only one protein.

Replication and particle assembly take place in the cytoplasm of the infected cell, where progeny virions are formed by a budding process from membranes of the rough endoplasmic reticulum. The virions acquire their lipid envelopes from the cells, excluding host-cell proteins in the process, and are subsequently transported through and accumulate in the Golgi complex and smooth-walled vesicles.

3.2   *Clinical and pathological aspects*

It is thought that 2–10 per cent of common colds are caused by coronaviruses, which, unlike rhinoviruses, can infect the lower respiratory tract. In a typical case there is an incubation period of 3 days followed by an unpleasant nasal discharge and malaise, lasting about a week. There is little or no fever, and coughs and sore throat are not common.

Pathogenesis

Replication is confined to the cells of the epithelium in the upper respiratory tract. Inflammation, oedema, and exudation occur in the tract for several days following destruction of cells by the virus.

Epidemiology

Coronavirus colds occur in the colder months of winter and early spring with sizeable outbreaks every 2–4 years. Antibody surveys show that most people have been infected at some time in their lives and it is thought that reinfections are quite common.

Laboratory diagnosis

Most routine clinical virology laboratories are not equipped to isolate coronaviruses, which replicate poorly in cell cultures and require organ cultures of human embryo trachea or nasal epithelium. In any event, the trivial nature of the infection does not call for routine diagnostic tests. For epidemiological purposes, paired acute and convalescent sera can be tested by HI for a rising titre of specific antibody.

4   RHINOVIRUSES

A landmark in the study of the virology of the common cold came in the 1960s when, after years of patient investigation, David Tyrrell and his colleagues at the MRC Common Cold Research Unit in Salisbury isolated viruses in a simple cell culture system. A chance observation that an alkaline pH in the culture medium favoured the growth of these viruses was the key to their success.

4.1   Properties of the viruses

Classification and general properties

Rhinoviruses (Greek *rhis*: nose) constitute a genus of the family *Picornaviridae* (see Chapter 14) and are small non-enveloped RNA viruses some 18–30 nm in diameter; they possess icosahedral symmetry and are morphologically similar to other picornaviruses (Fig. 14.2). By conventional neutralization tests they fall into more than 150 serotypes. However, the more recently applied techniques of RNA:RNA hybridization (Chapter 27) define different patterns of genetic relationships between the viruses and suggest that rather fewer than 100 different rhinoviruses exist. These genetic differences may have practical implications since some antiviral agents inhibit viruses only in particular groups.

In many properties, rhinoviruses resemble other members of the *Picornaviridae* (Chapter 14). They differ, however, in their inability to withstand acid conditions, and in their low optimum temperature for growth (33°C). The latter characteristic is undoubtedly an evolutionary adaptation to the comparatively cool environment of the nasal mucosa.

4.2   *Clinical and epidemiological aspects*

Replication is restricted to cells of the upper respiratory tract. The infection is spread by aerosol and by hand contact and the incubation period is 2–3 days. Inflammation, oedema, and copious exudation from the upper respiratory tract last for a few days. There are few serious sequelae except in chronic bronchitics, in whom attacks may be precipitated by infection with a common cold virus.

Rhinovirus colds occur throughout the year in all countries of the world. It is thought that several virus serotypes co-circulate for a year or so, to be displaced by a new group; one individual may experience two or three infections per annum. Children are more susceptible than adults and one of the hazards of living in a large family is the risk of contracting a cold from the younger members.

Antiviral chemotherapy or vaccines to prevent the common cold?

The potential antiviral effects of interferon were first established in volunteers infected with common cold viruses at the Salisbury experimental unit. This single experiment in 1973 used up almost all the world's supply of the new drug. Nowadays, with the arrival of recombinant DNA technology, a range of interferon molecules are available for clinical use. But even cloned and highly purified interferon is not without side-effects, a particularly unfortunate one being production of stuffy nose after intranasal use. The early promise of interferon as an effective antiviral compound has thus not yet been realized. A number of synthetic antiviral molecules, such as enviroxime and dichloroflavan (see Chapter 29), have good antiviral activity against certain rhinovirus serotypes in the laboratory, but only marginal effects have been noted in volunteer experiments.

Little progress has been made with vaccines because of the antigenic diversity of the rhinoviruses. Data from X-ray crystallography and nucleotide sequencing may help to delineate common amino-acid sequences among the capsid proteins of rhinoviruses, which could be used as peptide vaccines. But the effective control of the common cold lies very much in the future.

5 REMINDERS

◆ **Adenoviruses** are **dsDNA** viruses with an **icosahedral** structure embellished by a projecting fibre at each apex.

◆ Shipyard eye is an **epidemic keratoconjunctivitis** caused by serotype 8; many of the remaining 33 serotypes cause respiratory symptoms. Some adenoviruses cause outbreaks of **respiratory infection** in the winter, and others, **gastrointestinal illness**.

◆ **Coronaviruses** are **positive-strand RNA** viruses and two serotypes are known. They cause **upper respiratory tract infection**. These viruses are difficult to grow in the laboratory but readily spread and infect humans, causing up to 10 per cent of common colds.

◆ **Rhino** or **common cold viruses** are picornaviruses; there are 100–150 serotypes, explaining why repeated infections are not uncommon. The viruses cause **upper respiratory tract infection** at all times of the year throughout the world; they do not infect the lower respiratory tract.

◆ No universally effective vaccine or antiviral exists to counter the effects of these three groups of viruses although experiments are in progress with live attenuated adenovirus vaccines and with specific drugs against common cold viruses.

CHAPTER NINE

CHILDHOOD INFECTIONS CAUSED BY PARAMYXOVIRUSES

1 INTRODUCTION

The *Paramyxoviridae* cause a variety of diseases, predominantly involving the respiratory tract, in humans, animals, and birds. In humans, they include measles, respiratory infections caused by respiratory syncytial virus (RSV) and parainfluenza viruses, and the more innocuous salivary gland infection of mumps. These viruses, particularly RSV, cause fusion of infected cells with formation of multinucleated giant cells (syncytia).

2 PROPERTIES OF THE VIRUSES

2.1 Classification

Members of this family are enveloped, negative-stranded RNA viruses, 150–200 nm in diameter, with a nucleocapsid of helical symmetry. Within the family *Paramyxoviridae* three genera are recognized (Table 9.1); one distinguishing feature is the possession (*Paramyxovirus*) or absence (*Morbillivirus* and *Pneumovirus*) of neuraminidase.

Table 9.1 The *Paramyxoviridae*

Genus	Diseases of humans	Diseases of other species
Morbillivirus	Measles	Distemper in dogs; infections of the mucosae, gut, and respiratory tract in cattle, sheep, goats and pigs (rinderpest)
Pneumovirus	**Respiratory syncytial virus**	
	Upper and lower respiratory tract infections in infants, especially bronchiolitis	Respiratory infections in poultry and rodents
Paramyxovirus	Mumps	
	Parainfluenza viruses	
	Croup (types 1 and 2); bronchiolitis (type 3); upper respiratory tract infections (types 1–4)	Upper respiratory tract infections in various species, including Newcastle disease in poultry

2.2 *Morphology and structural proteins*

Because of the fragility of the lipoprotein envelope the viruses often appear distorted or disrupted in the electron microscope, with the nucleoprotein spewing out from inside the virion (Fig. 9.1). The structural polypeptides of the capsid include the HN (haemagglutinin–neuraminidase) and F (fusion) glycoproteins which form the surface spikes, and the M (matrix) protein. Three other proteins, together with the RNA, form the nucleoprotein core of the virion and possibly the RNA transcriptase possessed by all negative-stranded viruses.

The F protein, formed by proteolytic cleavage of a larger precursor polypeptide, is important, since it mediates the fusion of infected cells which then form the **syncytia** so characteristic of infections with this group of viruses.

2.3 *Genome*

The genome of the paramyxoviruses is a ssRNA molecule of some 15 000 nucleotides containing the genes coding for the six known virus-specific proteins.

2.4 *Replication*

The strategy of replication is that of a typical negative-stranded RNA virus (Chapter 2). The virion enters a cell by **fusion** of its envelope with the external plasma membrane of the cell and the genome is released immediately into the

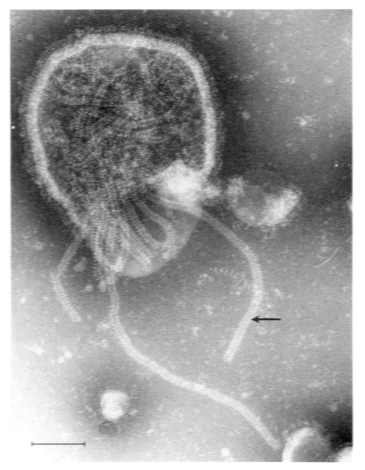

Fig. 9.1. Electron micrograph of a paramyxovirus. The virion has ruptured, spilling out coils of ribonucleoprotein (arrowed). (Courtesy of Dr David Hockley.) Scale bar, 100 nm.

cytoplasm. Viral proteins are synthesized from mRNAs which appear to be monocistronic, i.e. only one protein is encoded by a single mRNA species. The viruses are released from the infected cell by budding from the plasma membrane. Because of the tendency of infected cells to fuse, they may, under certain conditions, release hardly any new viruses into the extracellular environment; instead, the viruses may spread from cell to cell, behaving as typical 'creepers' (Chapter 4).

3 | CLINICAL AND PATHOLOGICAL ASPECTS

3.1 *Measles*

'Measles' derives from an Anglo-Saxon word, *maseles*, and so has obviously been with us for many centuries. Its Latin name, *morbilli*, is a diminutive of *morbus*, a disease, and thus signifies a minor illness. In temperate countries measles is indeed, although unpleasant, rarely serious, but in some tropical areas it is a killer.

Clinical features

The period from infection to appearance of the rash is fairly constant at about 2 weeks but may be less by a few days. Before the rash there is a **prodromal stage lasting 2 or 3 days**, with running eyes and nose, cough, and moderate fever. At this time, careful examination of the buccal mucous membrane adjacent to the molar teeth may reveal **Koplik spots**, which resemble grains of salt just beneath the mucosa. They may be many or few in number, but when present, are pathognomonic of measles.

The **rash** first appears on the face and spreads to the trunk and limbs within the next 2 days. It is a dull-red, blotchy, **maculopapular** eruption, usually characteristic enough to have its own designation, 'morbilliform' (see Fig. 4.1(a)). If severe, it may have a purpuric appearance.

Although the rash is one of the most obvious physical signs, it is but one manifestation of a generalized infection. The temperature rises sharply to about 40°C, and in every case there is evidence of **bronchitis** and **pneumonitis**, with cough and 'crackles' in the chest. Occasionally, diarrhoea in the early stages indicates inflammatory lesions in the gut.

Within 2–3 days of onset, the rash starts to fade, the temperature subsides, and the child feels better; recovery is usually uneventful.

Complications

Along with the fever and rash, there are widespread lesions of various body systems (Table 9.2). Although they are listed as complications, it is arguable whether most of them should be referred to as such, since they are all part of the generalized infection and are thus present in most cases; whether they surface in the form of signs and symptoms depends on their severity. The exceptions are those in which the immune response of the host is abnormal. They include postexposure encephalitis (about 1 per 1000 cases), subacute sclerosing panencephalitis (about 1 per million cases), and giant-cell pneumonia, a life-threatening infection seen occasionally in immunodeficient children (Fig. 26.2).

Pathogenesis and pathology

Infection is acquired via **droplets** that enter the respiratory tract. The pathogenesis is that of an acute generalized infection (Chapter 4), the secondary viraemic phase corresponding to the height of the fever.

The rash is due not to a direct cytopathic effect of virus but to a reaction by cytotoxic T cells against viral antigen appearing in the skin cells. In addition, antigen–antibody complexes form on the capillary endothelium with consequent cell damage, vasodilation, and leakage of plasma. The rash

Table 9.2 Complications of measles

Site	Complication
Respiratory system	Croup (in prodromal stage); bronchitis; giant-cell pneumonia (in immunocompromised patients)
Eye	Conjunctivitis; corneal ulceration (rare)
Ear	Otitis media; possible secondary bacterial infection
Gut	Enteritis with diarrhoea
Central nervous system	Febrile convulsions (acute phase); post-exposure encephalitis (rare) (Chapter 24); subacute sclerosing panencephalitis (very rare) (Chapter 24)

is thus a clear sign that a satisfactory immune response is in progress, and that recovery is on the way; conversely, its failure to appear, e.g. in immunodeficient patients, is a bad prognostic sign.

During the prodromal and acute phases, virus is **shed** in body fluids, including **respiratory secretions**; it replicates in **leukocytes**, causing **leukopenia**. During the acute phase, virus-infected giant cells containing up to 100 nuclei have been found in the pharynx and tonsils, skin, respiratory epithelium, lymph nodes, and Peyer's patches. The virus is widespread in the skin but disappears quickly with the onset of the rash and the appearance of circulating antibody.

Immune response

IgM, IgA, and IgG antibodies appear at the same time as the rash, the first two disappearing within a month or so, the latter persisting for life. During the acute phase, replication of virus within monocytes and other white cells depresses cell-mediated responses to other antigens, although cytotoxic T lymphocytes specifically directed against measles-infected cells are important in the recovery process.

Epidemiology

Measles virus is highly infectious and has a world-wide distribution. It becomes **endemic** only in countries with populations large enough to provide a continuing supply of susceptible children; in small isolated communities, it dies out until a fresh importation causes a major epidemic (see Chapter 7). In countries with high standards of living but poor immunization rates, endemic infection is punctuated every 2 to 3 years by epidemic infections that attack predominantly 3–5 year-olds, in whom the illness is relatively mild. In Third World countries, on the other hand, measles has a high incidence in infants less than 2 years old and is much more severe, with unusual clinical features such as blindness, and a high fatality rate (see 'Measles in developing countries').

Laboratory diagnosis

In day-to-day practice, laboratory confirmation of the clinical diagnosis is not needed. In atypical cases, usually seen in hospitals, **serological** diagnosis is made by **complement-fixation** or **haemagglutination-inhibition** tests for a fourfold or greater rise in antibody titre, or for the presence of specific IgM antibody.

Prophylaxis

Passive protection with normal human immunoglobulin (NHIG) in a dose sufficient to modify but not completely prevent measles is valuable for protecting debilitated or immunodeficient children who have been exposed to infection.

Measles vaccines. Early trials of formalin-inactivated vaccine met with failure because, on subsequent exposure, some of the recipients developed unusually severe measles; the probable reason was that the formalin had damaged the fusion protein in the vaccine, so that antibody against this important component was not stimulated. Modern **attenuated vaccines** are, however, very effective, and ideally are given in combination with mumps and rubella vaccines (**MMR vaccine**). This preparation is now in use in the UK; it is given at **13–15 months** of age, by which time there is no risk of neutralization by maternal antibodies. The preparation of vaccine is simplified by the fact that there is only one serotype.

In the USA, an attempt is being made to eradicate measles by achieving vaccination rates approaching 100 per cent. As part of the campaign, children are not admitted to state schools without a certificate of vaccination. These efforts are now bearing fruit: measles deaths have declined from around 400 *per annum* in 1960 to fewer than 10 at present. Similarly, cases of measles encephalitis reported annually have fallen from about 300 to single figures. The economic saving in medical care and hospital beds is enormous.

Measles in developing countries

Two host factors seem of primary importance in explaining why measles is much more severe in developing countries than in the United States and Europe: (1) on average, children acquire the infection at a younger age; and (2) they are less well nourished. During the Nigerian civil war, when there was severe famine, at least 15 per cent of children infected with measles died.

In these children, measles differs considerably from the relatively benign illness described above, and more nearly resembles the disease as it existed in the Europe of 100 years ago. The **rash** is more severe and exfoliates extensively, exposing large areas of skin to bacterial invasion. **Stomatitis** interferes with eating and drinking. **Vitamin A deficiency** and measles infection combine to cause **corneal ulceration** and blindness. The virus is cleared less rapidly than in well nourished people; this results (1) in more damage to leukocytes with **depression of cell-mediated immunity** and **increased susceptibility to bacterial infections** and (2) in prolongation of the period of virus shedding by a week or more. **Bronchopneumonia** and persistent **diarrhoea** with protein-losing enteropathy often result in death, mortality rates being of the order of 10 per cent.

Measles vaccine in developing countries

The heat lability of measles vaccine complicates its distribution in tropical climates. The freeze-dried vaccine has to be held at 2–8°C until minutes before injection and, unless this temperature range is maintained, the vaccine rapidly loses its potency. The other problem in providing effective immunization for children in developing countries is reaching them during the brief interval between the loss of maternal antibodies and the acquisition of natural disease (the so-called 'window of opportunity'). This may be solved by the use of more concentrated ('high-dose') vaccine, which is not swamped by maternal antibody and is effective in babies less than 9 months old.

3.2   Respiratory syncytial virus infections

This virus — RSV for short — is highly contagious and causes sharp outbreaks of respiratory disease throughout the world, particularly in infants.

Clinical aspects

The illness often starts like a common cold but within 24 h the baby may be acutely ill with cyanosis and respiratory distress. Typically, there is **bronchiolitis**, with or without involvement of the lung parenchyma causing **pneumonitis**. There is some evidence that RSV infection in infancy may cause long-term respiratory problems. Severe illness carries a significant risk of death.

Pathogenesis and pathology

The incubation period is about 5 days. There is a ***necrotizing bronchiolitis*** in which partial blocking of the bronchioles leads to the collapse of areas of lung. Peribronchial infiltration may spread to give widespread interstitial pneumonitis. Inapparent infections also occur.

Immune response

Since infants as young as 6 weeks are often infected, maternal antibody does not seem to provide protection for very long after birth. RSV induces both antibody and cell-mediated responses; impairment of the latter in immunocompromised patients may lead to **persistent infection**. The **interferon** response is noticeably worse than in other paramyxovirus infections. It has been suggested that harmful immune responses, e.g. formation of antigen–antibody complexes, plays a part in pathogenesis, but this has not been proved. In later life, **reinfections** are comparatively frequent, suggesting that first infections do not always induce long-term immunity. Antigenic variation between strains of RSV may also contribute to reinfections.

Epidemiology

RSV infections are transmitted by infected **respiratory secretions**, in this case, not mainly by small droplets but by **contamination of hands** or **fomites** such as bedding. In temperate climates, they occur annually in epidemic form during the winter months but, in the tropics, the incidence may be highest during the summer months or rainy season. Babies aged from 6 weeks to 6 months are predominantly affected; indeed, RSV is the most important respiratory pathogen in young infants.

During epidemic periods, spread within hospitals, crèches, and day nurseries (**nosocomial infections**) often takes place, facilitated by close personal contacts and the liability of those infected to shed virus for up to 3 weeks after the acute phase. Adults may also acquire RSV, particularly the elderly, in whom it may exacerbate existing bronchitis.

Laboratory diagnosis

The virus can be isolated in continuous lines of human cells, e.g. HeLa, in which cytopathic effects with formation of syncytia appear after 2–10 days. The method of choice, however, is demonstration by **indirect immunofluorescence** of viral antigen in cells from a nasopharyngeal washing, which gives an immediate result.

Prevention

Infants given experimental inactivated RSV vaccines developed antibody, but contracted more severe disease than non-immunized controls when exposed to infection, a result similar to that obtained with inactivated measles vaccine and probably due to a similar mechanism (Section 3.1). Attempts to develop attenuated vaccines have so far met with little success; the use of recombinant DNA technology for making RSV vaccine is now being studied.

In the absence of a vaccine, limitation of spread within paediatric units, nurseries, and the like depends on good hygienic practice such as handwashing, covering of the mouth when coughing or sneezing, and careful disposal of paper handkerchiefs, etc.

Treatment

Small-scale trials of ribavirin (Chapter 29), given by continuous aerosol inhalation to babies with serious RSV infections met with some success, but not enough to make it a routine treatment. At present, ribavirin should perhaps be reserved for infants with pre-existing cardiopulmonary disease, who readily succumb to chest infections.

3.3 *Mumps*

The name probably originates from an old word meaning 'to mope', an apt description for the miserable child afflicted by this common illness. Mumps was one of the first infections to be recognized and was described by Hippocrates as early as the fifth century BC. Robert Hamilton in *An Account of a Distemper by the Common People of England Vulgarly Called the Mumps*, noted in 1790 that 'The catastrophe was dreadful: for the swelled testicles subsided suddenly the next day, the patient was seized with a most frantic delirium, the nervous system was shattered with strong convulsions, and he died raving mad the third day after.' Fear of orchitis and consequent sterility, in reality greatly exaggerated, explains the continued interest in immunization against what is otherwise a comparatively benign disease.

Clinical aspects

The onset is marked by malaise and fever, followed within 24 h by a painful enlargement of one or both parotid glands; the other salivary glands are less often affected. In most cases, the swelling subsides with a few days and recovery is uneventful. Inapparent infections are common.

Complications

Orchitis develops in about 20 per cent of males who contract mumps after the age of puberty; it may develop in the absence of preceding parotitis. Typically, there is pain and swelling of one or both testicles 4–5 days after the onset of parotitis. The pain is often severe enough to demand strong analgesics. There is often an accompanying general reaction with high temperature and headache. Symptoms tend to subside after 3–6 days. Although some degree of testicular atrophy follows in about 30 per cent of cases, sterility following mumps orchitis is rare.

Inflammation of the **ovary** (oophoritis) and **pancreas** has been reported, but these complications do not seem to have serious long-term effects.

Central nervous system. The incidence of 'aseptic' **meningitis** is higher after mumps than after any other acute viral infection of childhood. Rates of 0.3–8 per cent have been reported in the USA. This complication almost always resolves without sequelae. **Postexposure encephalitis** is, however, more serious and carries an appreciable mortality. These syndromes are described in Chapter 24.

Some degree of **deafness** is a residual complication in a small percentage of cases.

Pathogenesis

The infection is spread in saliva and secretions from the respiratory tract, and is acquired by the **respiratory route**; the incubation period is 14–21 days. Viraemia during the acute phase is followed by generalized spread to various organs, including the parotid gland. Virus is shed for several days before and after the first symptoms, not only from the respiratory tract but also in the urine.

Immune response

The appearance of specific IgM, IgA, and IgG antibodies follows the sequence usual for these acute viral infections. Cell-mediated immunity is probably important in the recovery process.

Epidemiology

Mumps has a world-wide distribution and mainly affects those aged less than 15 years. In temperate climates, the incidence is highest in winter, but there is no seasonal variation in tropical countries. Mumps is highly infectious: outbreaks in institutions are common and there have been a number among military recruits in barracks.

Laboratory diagnosis

Virus can be isolated in various cell lines and identified by haemadsorption or haemagglutination-inhibition (Chapter 27). Serological tests are, however, more widely used; they include **ELISA** tests for **specific IgM** and **complement fixation tests**. The latter is used to test for **antibody to the S (soluble) antigen** which appears early and is transient; its presence thus indicates an acute or recent infection. Antibody to the V (viral surface antigen) persists for much longer, and its demonstration, in the absence of anti-S, only indicates past infection.

Prevention

Live mumps vaccines prepared either from the Jeryl Lynn or Urabe attenuated strains are available. They are both highly effective, but there is an appreciable incidence of post-vaccination meningitis with the latter, which has now been abandoned in the UK. In the UK and USA mumps vaccine is combined with measles and rubella vaccines (MMR). Where uptake rates are high, there have been very substantial reductions in the number of cases of mumps and its complications.

3.4   Parainfluenza viruses types 1–4

These four respiratory viruses are world-wide in their distribution and cause respiratory disease in all age groups, but predominantly in children (Table 9.1). Reinfections are common, although they may often be subclinical. These subclinical infections maintain a huge reservoir of infective virus. In a unique study at the Antarctic base at the South Pole, parainfluenza viruses were isolated continuously, although the personnel were completely cut off from the outside world for several months during the winter.

The viruses are transmitted by airborne droplets and spread is rapid among children in institutions.

Parainfluenza viruses cause up to one-third of all respiratory tract infections and nearly one-half of respiratory infections in pre-school children and infants.

Types 1 and 2 are most often associated with laryngotracheobronchitis (croup), boys being affected more often than girls; type 3 usually causes infection of the lower respiratory tract, e.g. bronchiolitis and pneumonia.

What little is known of the **pathogenesis** and **immune response** suggests similarities with those of RSV infections. The **epidemiology** is also broadly similar to that of other respiratory infections due to paramyxoviruses. The methods of **laboratory diagnosis** are similar to those for RSV infections, in particular, the use of **indirect immunofluorescence** for detecting antigen in nasopharyngeal washings.

Prevention

Inactivated vaccines met with the same fate as those prepared for measles and RSV, probably for the same reasons. The successful use of an attenuated vaccine for the related and economically important virus infection of poultry, Newcastle disease, suggests that similar vaccines may eventually be prepared for use in humans.

4 | REMINDERS

- ◆ The family *Paramyxoviridae* contains three genera: *Morbillivirus*, *Pneumovirus*, and *Paramyxovirus*. The viruses are about 200 nm in diameter, of **helical** symmetry, and possess a ssRNA genome.

- ◆ All the viruses code for a fusion protein (F), which causes adjacent infected cells to fuse and form **multinucleate giant cells (syncytia)**.

- ◆ Morbilliviruses cause **measles** in humans and other generalized infections, including canine distemper, in animals. Measles is an acute febrile illness of childhood associated with a characteristic **maculopapular rash**, which results from cell-mediated cytotoxicity and an interaction of virus in capillary endothelium with newly formed antibody. Complications include **encephalitis**, **subacute sclerosing panencephalitis** (very rare), and, in patients with defective immunity, **giant-cell pneumonia**. **Severe forms** of measles with high mortality rates are seen in some developing countries and are associated with **malnutrition**.

- ◆ **Passive protection** with normal human immunoglobulin is used to protect immunodeficient children exposed to infection. **Active immunization** with attenuated measles vaccine, alone or in combination with mumps and rubella vaccines (MMR), is highly effective.

◆ **Respiratory syncytial virus (RSV)** affects the **lower** respiratory tract in infants; it may cause **necrotizing bronchiolitis** and **pneumonitis**. RSV is spread more by contamination of hands and fomites than by droplets and, in the absence of an effective vaccine, prevention in institutions depends on good hygiene. For rapid diagnosis, the best method is demonstration by **indirect immunofluorescence** of viral antigen in cells from nasopharyngeal washings.

◆ **Mumps** is a generalized infection of childhood, in which the salivary glands, especially the parotids, are attacked. It is a comparatively benign illness. In 20 per cent of infected adolescent and adult males **orchitis** develops, but rarely results in sterility. Other complications include **'aseptic' meningitis**, **postexposure encephalitis**, and **deafness**.

◆ Laboratory diagnosis depends on isolation of the virus, or, preferably, serological tests. A positive complement-fixation test for **antibody to soluble (S) antigen** indicates current or recent infection.

◆ An effective **attenuated vaccine** is available.

◆ **Parainfluenza viruses** types 1–4 are a major cause of respiratory tract infections, especially in infants and children. Types 1 and 2 are particularly liable to cause **croup**, and type 3, **bronchiolitis**. All four serotypes also cause upper respiratory infections. The pathogenesis, epidemiology, and methods of diagnosis are similar to those of RSV.

CHAPTER TEN

ORTHOMYXOVIRUSES : INFLUENZA

1 INTRODUCTION

Influenza has long been with us; indeed, the name itself refers to the ancient belief that it was caused by a malign and supernatural influence. Known in the sixteenth century as 'the newe Acquayntance', influenza still causes major outbreaks of acute respiratory infection. It has indeed been described as 'the last great uncontrolled plague of mankind' and in this chapter we shall show how the property of causing epidemics and even pandemics is directly related to the ability of the causal viruses to undergo antigenic variation and thus evade their hosts' immune defences.

2 PROPERTIES OF THE ORTHOMYXOVIRUSES

2.1 Classification

Although laymen (and some doctors who should know better) refer to many incapacitating respiratory infections as 'flu', true influenza is caused by the small family of the *Orthomyxoviridae*. *Myxo* derives from the Greek for mucus and refers to the ability of these viruses to attach to mucoproteins on the cell surface; *ortho* means true or regular, as in orthodox, and distinguishes these viruses from the paramyxoviruses (Chapter 9). There are **three species, A, B, and C**, distinguished serologically on the basis of their matrix (M) and nucleoprotein (NP) antigens. Influenza C differs significantly from A and B and is of much less importance in infections of humans; the descriptions that follow relate only to the A and B viruses.

Since 1971, influenza A viruses have been designated on the basis of the antigenic relationships of the haemagglutinin (HA) and neuraminidase (NA) proteins. HA and NA antigens deriving from animals and birds were at first given appropriate letters, e.g. Hsw for the haemagglutinin of a swine-type virus or Nav for a neuraminidase of avian origin. You may see these designations in older publications, but since 1980 the antigens have been given simple sequential numbers, H 1–13 and N 1–9. Of these, only viruses with H1, 2, or 3 and N1 or 2 are known to infect humans. You may also see, for example, in literature describing the latest influenza vaccine, the designations of the individual strains from which it was prepared: these follow the pattern A/Mississippi/1/87 (H3N2), (where A is for influenza A, followed by the place where it was isolated, the laboratory number, the year of isolation, and the H and N subtypes). Type B strains are designated on the same system, but without H and N numbers since antigenic shift in these viruses has so far not been observed.

2.2   *Morphology*

The virions are 100–200 nm in diameter and are more or less spherical. Laboratory strains however, may be very pleomorphic, sometimes with filamentous forms as long as 1000 nm. The symmetry is helical.

The lipid envelope is covered with about 500 projecting spikes, which can be clearly seen under the electron microscope (Fig. 10.1). About 80 per cent of them are haemagglutinin antigen and the remainder are another antigen, neuraminidase. Within the lipid envelope is a layer of matrix protein which itself encloses the RNA genome of the virus. The RNA is protected by an associated protein called the nucleoprotein (Fig. 1.8).

2.3   *Genome and polypeptides*

The eight discrete fragments of the RNA genome (Fig. 10.2) are complexed with protein to form a ribonucleoprotein arranged in a helix. These RNA segments function as individual genes and code for the various non-structural and structural polypeptides; the important haemagglutinin and neuraminidase proteins are coded for by segments 4 and 6, respectively. The **haemagglutinin (HA)** is a rod-shaped glycoprotein with a triangular cross-section (Fig. 10.2). This antigen was first identified by its ability to agglutinate erythrocytes, hence its name, but it is now apparent that it also has important roles in the

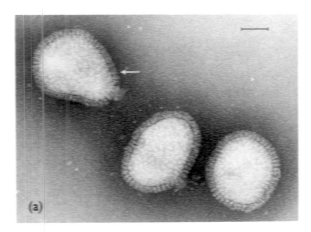

attachment and entry of virus to the cells of the host and in determining virulence. For the virion to be infective, HA must be cleaved into two molecules, HA1 and HA2; this is effected by a host-cell protease after the HA has been synthesized on the host-cell ribosomes.

The **neuraminidase (NA)** is, as its name implies, an enzyme that can destroy neuraminic (sialic) acid, a component of the specific cell receptors for these viruses. It may play some part in attachment of virus, but its main function seems to be connected with release of new virus from cells. Like HA, it is a glycoprotein, but the NA spikes look like mushrooms rather than rods (Fig, 10.2).

The HA and NA molecules penetrate through the lipid bilayer of the virion and probably make contact with the underlying layer of matrix (M) protein, which both contributes to the integrity of the virion and is involved in assembly of the virus at the cell plasma membrane before budding and release. Inside the matrix shell are the nucleoprotein and an RNA transcriptase that is essential for replication (Section 2.4).

Fig. 10.1. Electron micrographs of influenza viruses. **(a)** Negatively stained virions. The surface haemagglutinin and neuraminidase antigens are arrowed. Scale bar, 50 nm. **(b)** Scanning electron micrograph of virus particles budding from the cell surface. Many thousands are present on an infected cell. (Courtesy of Dr David Hockley.) Scale bar, 500 nm.

2.4   *Replication*

After attachment to the specific receptors on the cell membrane, virions are taken into the cell by endocytosis and then transported to vacuoles (endosomes), where the acid pH induces a change in the configuration of the HA. This brings a special set of amino acids, the 'fusion sequence', in contact with the lipid of the vacuole wall. Fusion with the vacuole triggers uncoating of the virus, after which its RNA is transported to the host-cell nucleus.

Orthomyxovirus RNA is negative-strand, so that the first step in replication must be its transcription by the virus's own RNA polymerase to a complementary positive strand that can function as mRNA. Thereafter, the general pattern is that described for negative-strand RNA viruses in Chapter 2.

The new virions are assembled at the host-cell surface membrane and released by a process of budding in which both HA and NA are involved.

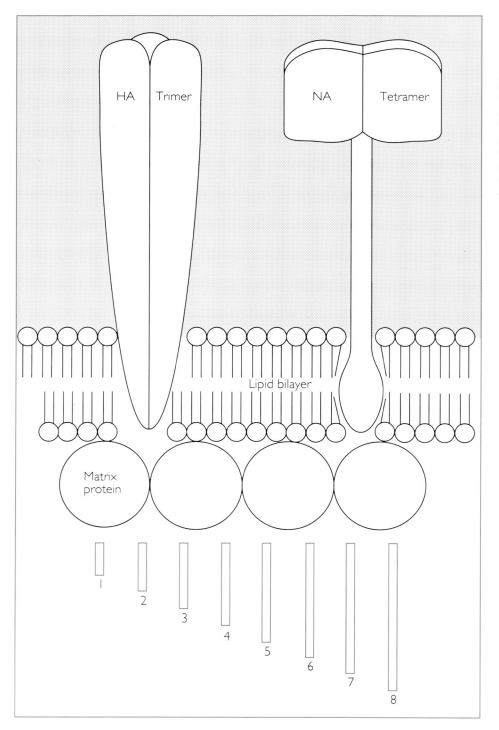

Fig. 10.2. The haemagglutinin (HA) and neuraminidase (NA) antigens of influenza. The triangular HA and mushroom-shaped NA spikes protrude from the viral lipid in a ratio of 5:1. A layer of matrix protein underlies the lipid membrane and encloses the eight segments of viral RNA.

2.5   *Genetic variation in influenza viruses*

Gene reassortment: antigenic shift

▪▪▪▪▪▪▪   Influenza A viruses—but not B or C—readily undergo gene 'swapping' or reassortment (Fig. 10.3), so that, in a cell infected simultaneously with two different viruses, the progeny virions may contain mixtures of each parent's genes. Add this property to the ability of influenza A virus to infect animals and birds that often live in close association with man, and we have a situation in which double infections with viruses of human and non-human origin result at unpredictable intervals in the formation of new strains with genetic compositions differing from those in general circulation. This reassortment of genes, known as antigenic shift, can, of course, also take place between two strains of human origin.

If these exchanges involve the genes coding for HA, NA, or both, it is clear that the new 'reassortant' strains will have a selective advantage in that the herd immunity of the population, which is largely mediated by antibodies to these antigens, will be relatively ineffective. Thus antigenic shift can result in widespread epidemics of influenza A (Fig. 10.4).

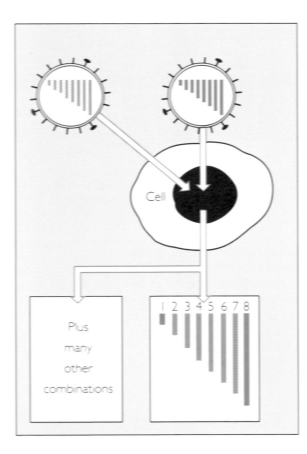

Fig. 10.3. Genetic reassortment with influenza A virus. Infection of a single cell by two different influenza A viruses results in their replication and the subsequent exchange of any of the eight genes of each parent.

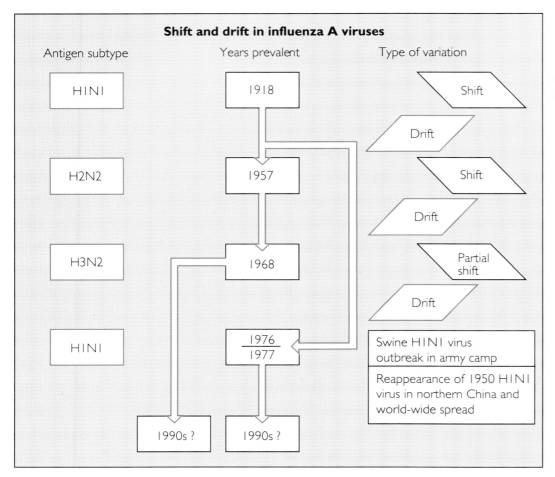

Fig. 10.4. Antigenic shift and drift. For explanation, see text.

Antigenic drift

RNA viruses tend to have high mutation rates—say 10 000 times greater than that of human or viral DNA—and this is true of all the influenza viruses. These mutations also give rise to changes in the genetic material, and hence in the viral polypeptides, which undergo two or three amino-acid substitutions each year; but being slowly progressive and cumulative, they are not as dramatic as those caused by antigenic shift. The capacity of the 'drifted' strains to spread in the community depends on whether the HA and NA antigens are affected; but since the changes are less pronounced than in antigenic shift, these strains tend to cause localized outbreaks rather than big epidemics.

Note that both influenza A and B undergo antigenic drift.

3 CLINICAL AND PATHOLOGICAL ASPECTS

3.1 *Epidemic influenza*

The genetic lability of influenza viruses make them capricious and formidable enemies to world public health. The interchange of the A viruses between birds, man, and other animals is well documented and can, in fact, be reproduced under laboratory conditions. Recently, avian strains were isolated from seals dying in large numbers on the eastern seaboard of the USA and viruses with characteristics of horse and swine influenza A viruses have turned up in humans from time to time. The wide range of animal hosts is depressing to contemplate, since it implies that it will never be possible to eradicate the disease completely in humans.

This promiscuous behaviour of influenza A is well suited to the emergence, at intervals of a decade or more, of 'shifted' strains, which, together with continual antigenic drift affecting both A and B viruses, confers flexibility in evading the action of host antibodies or cellular immunity; furthermore, a strain may lurk in an animal reservoir — or possibly even in humans — for many years before emerging to cause outbreaks in populations whose herd immunity to that particular virus has waned with the passage of time. It is clear that this quality of evasiveness confers upon influenza great survival potential and we shall now see how it gives rise to large and sometimes devastating epidemics.

Figure 10.4 shows the relationship between antigenic drift, shift, and epidemics of influenza A. Note that, since influenza virus was not isolated until 1933, data for previous years are based on antibody studies of the populations then at risk.

The pandemic of 1818–19 was especially terrible in its effects in Europe; world-wide, it killed about 20 million people, far more than lost their lives in the whole period of the First World War. The infection spread rapidly both in the armies in the field and in civilians, and malnutrition following 4 years of war may have contributed to the high mortality in the latter population. Although then known as 'Spanish influenza', it is likely that the outbreak started in Asia; this is certainly true of the pandemic that occurred nearly 30 years later, when a strain differing completely in both HA nd NA antigens appeared. This H2N2 virus spread over the entire globe within a few months of its first isolation in China. In 1968 there was another pandemic, again originating in the Far East; the virus, first isolated in Hong Kong, had now undergone a partial shift that affected only the HA. Perhaps for this reason, the illnesses that followed were milder than in the previous pandemics, there being some herd immunity to the NA component.

In 1976 there was considerable alarm in the USA when the dreaded swine-type H1N1 influenza A virus appeared in a military barracks. This was a drifted variant related to the 1918 virus; it could have arisen again by

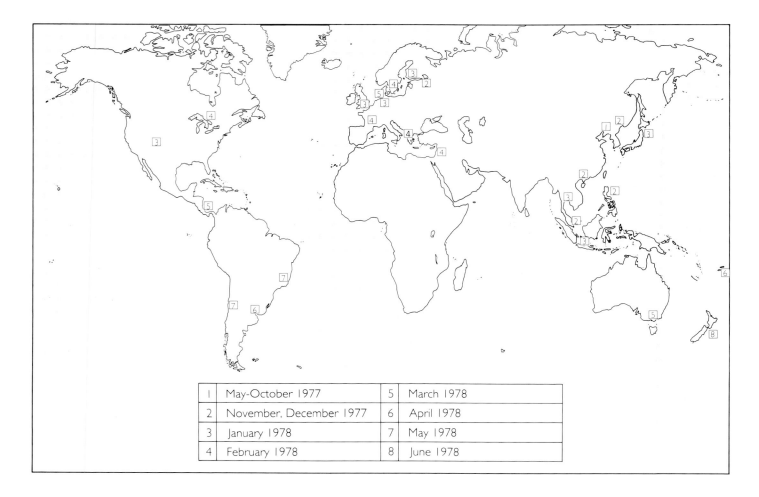

1	May-October 1977	5	March 1978
2	November, December 1977	6	April 1978
3	January 1978	7	May 1978
4	February 1978	8	June 1978

genetic reassortment, but it is more likely that the strain circulated for many years in an animal reservoir such as pigs before surfacing in man. Whatever its origin, its reappearance prompted the United States health authorities to undertake a massive immunization campaign involving 40 million doses of vaccine; but in the event—and not as a result of this programme—the feared pandemic never materialized. Since 1977 the H3N2 and H1N1 strains have circulated side by side (Fig. 10.5).

In between these massive epidemics at intervals measured by decades, there are annual outbreaks, usually during the colder part of the year, mostly caused by drifted strains of influenza A; influenza B causes fewer outbreaks, which tend to be localized and confined to young children. Influenza C infects most people during childhood, more often than not without causing significant illness.

Unlike shift, which, immunologically speaking, catches large populations unaware, drift is in a sense the result of the antibody response of the community. Point mutations, deletions, or insertions in a given strain lead to minor changes in the HA or NA antigens and give it a selective advantage

Fig. 10.5. Global spread of influenza A. The H1N1 virus was first detected in 1977 in China and rapidly spread into the USSR (now CIS), Europe, and then the USA and Oceania. (Modified, with permission, from WHO Weekly Epidemiological Record.)

over the prevalent antibody; this strain then becomes the dominant one until it in turn is replaced by another drifted variant of the same subtype. One can think of the populations of viruses and hosts as enemies constantly probing for chinks in each others' defences.

3.2 The origin of shifted strains of influenza A

Why do pandemic strains of influenza A seem to originate in southern China? The answer may lie in the close association of humans with domestic animals and birds. Largely unremarked except by farmers and veterinarians, influenza A viruses are constantly circulating in pigs, horses, and birds, including poultry, and it is a reasonable assumption that, in areas of very intensive small-scale farming, the chances of interchange of viruses between man and other species—and hence of genetic reassortment—are considerable. For example, influenza viruses can be recovered both from duck faeces and from pond water, and exchanges of viruses between swine and humans in both directions are also well recorded. There is, of course, no proof that epidemics starting in Asia originate in this way, but it is an attractive hypothesis. Another factor that may be important is the very high population density in that area, in which lives half the world's population. A less accepted hypothesis is that influenza A viruses continually recycle in humans and 'new' pandemic viruses are really old viruses re-emerging from a previously infected individual.

3.3 Clinical features

After an incubation period of 2–3 days there is usually a very abrupt onset with shivering, malaise, headache, and aching in the limbs and back. Characteristically, the patient is prostrated and has to take to bed. The temperature rises rapidly to around 39°C.

On examination, the cheeks may be flushed and the conjunctivae infected; there may be some pharyngitis. There are few if any physical signs in the chest and the X-ray is normal unless bacterial infection supervenes. The white cell count is either normal or shows a **mild leukopenia**, unless secondary bacterial infection causes a sharp rise in the neutrophil count.

Course of the illness
Fortunately, influenza is usually short-lived; the temperature drops by day 3 and the patient is out of bed by the end of the first week. In older people, however, recovery may take much longer, with persistent weakness and lassitude. In general, the severity of influenza is proportional to age.

Complications
Apart from secondary bacterial infection there are few complications, but one rare condition, Reye's syndrome, is sometimes associated with influenza in

children, often of the B type. The taking of aspirin has also been implicated in the causation of this syndrome, which involves encephalopathy with fatty degeneration of the liver and other viscera; it is often fatal (see also Chapter 24).

3.4   *Pathogenesis*

Infection is acquired by the respiratory route and is usually an infection of the upper respiratory tract. Virus multiplies in the epithelium and destroys the cilia, which are an important element in the defence of the respiratory system; this is followed by transient viraemia which, allied with the production of interferon, may account for the severe malaise.

Viral infection of the lower respiratory tract in the form of influenzal pneumonia sometimes occurs, presenting as an overwhelming toxaemia with a high mortality. Pneumonia is, however, more often due to secondary infection with bacteria. Of these, *Staphylococcus aureus* is both the most frequent and the most dangerous, leading to respiratory distress, cyanosis, and collapse within 2–3 days of the onset of infection. The idea that cell damage caused by the viral infection opens the door to the staphylococcus has been somewhat shaken by recent research: certain strains of *S. aureus* produce protease that cleaves HA into HA1 and HA2 (see Section 2.3) and thus enhances the infectivity of some strains of influenza, so that the synergism between bacterium and virus may in fact work the other way round. Less commonly than staphylococcal infection, and later in the course of the virus infection, there is secondary invasion by *H. influenzae*, pneumococci, or haemolytic streptococci. Such complications are particularly liable to occur in patients with pre-existing chronic bronchitis or heart disease.

3.5   *Immune response*

We have seen that immunity to infection is mediated by antibodies, the antihaemagglutinin being the most important in this respect, since it prevents virus from attaching to cells in the first place, whereas anti-neuraminidase probably prevents the release from the host cells. Antibody of the IgA class is probably important in preventing infection, acting as it does at the mucous surfaces of the respiratory tract.

If infection does occur, various arms of the immune system come into play to effect recovery. IgG antibodies in the blood limit viraemia and interferon also probably helps in this respect. Non-specific inhibitors of influenza virus are also known to circulate: they are not antibodies and their role in prevention of infection and recovery is not clear.

In addition to these humoral factors, cell-mediated immunity in the form of cytotoxic T cells and alveolar macrophages may also play a part in recovery.

3.6 *Laboratory diagnosis*

In general practice, the diagnosis is made on the evidence of the characteristic clinical picture backed up by the knowledge that an outbreak is in progress. Sometimes, however, particularly in hospital patients, a virologically confirmed diagnosis of an individual is desirable; it is also necessary for laboratories to isolate and identify the prevalent strains both for epidemiological purposes and for vaccine production.

Virus isolation

Virus can be readily recovered from throat swabs or nasal washes during the first few days of illness. The swab is placed in a buffered transport medium containing a little protein, e.g. bovine serum albumin, to stabilize the virus; the medium is inoculated as soon as possible into chick embryos or cell cultures, in which its presence can be detected by haemagglutination or haemadsorption. The species and subtype are then identified by further tests, including haemagglutination-inhibition and radial diffusion (Chapter 27).

Rapid diagnosis

If isolation of influenza virus in culture for further examination is not essential, it can be rapidly identified by immunofluorescence staining of cells in nasopharyngeal aspirates, a useful method in hospital practice. This method is however likely to be replaced by automated ELISA-type tests.

At one time there was little practical purpose in using rapid tests, but the advent of anti-influenza drugs has made their development of some importance: we shall now see what is being done in the way of specific chemotherapy and chemoprophylaxis.

4 PROPHYLAXIS

4.1 *Chemoprophylaxis*

Influenza A viruses — but not B or C — are inhibited by amantadine, a primary amine, and rimantadine, a methylated derivative (Chapter 29). In the UK amantadine has the proprietary name 'Symmetrel'; rimantadine is at present most widely used in the CIS (formerly USSR) and Eastern Europe. When discussing replication, we mentioned the role of the acid pH of the lysosomes in triggering the process of fusion and release of viral RNA into the cell nucleus; the mode of action of amantadine is not fully understood, but part of the story may involve interference with fusion and uncoating of the virus by raising the lysosomal pH.

The role of amantadine in therapy is difficult to evaluate because the acute phase of influenza during which virus is replicating is so short. Nevertheless, several controlled trials have shown that both amantadine and rimantadine, if given during the first day of illness, shorten the average duration of pyrexia. It would be useful to evaluate the use of this drug for patients who also suffer from cardiovascular or chronic respiratory disease, since they are at greatest risk from influenza.

4.2   *Influenza vaccines*

Immunization, rather than chemoprophylaxis, remains the method of choice for preventing both influenza A and B. Even so, immunization poses a particular problem: every time a new strain of influenza A appears, the rapid production of large quantities of virus with the required antigenic characteristics, together with the need for routine tests of safety and efficacy, limits the amount of vaccine available.

Since there is never enough vaccine to protect the whole population, who should be immunized? There are three main categories:

◆   **individuals at special risk**. These include old or debilitated people and those with chronic heart, respiratory, renal, or endocrine disease;

◆   **people in closed institutions** in which attack rates may be high;

◆   **groups in community service**, such as health-care staff and police, who may need protection against wholesale sickness at times of major epidemics.

Nowadays, there is a fourth category, people 'who wish to be given some protection against influenza'. This is possible because only a minority of the population actually ask to be vaccinated. In the USA it is estimated that about 5 per cent of the population receive vaccine.

Live or killed vaccine?

Like other live-virus vaccines, an attenuated influenza preparation has the advantage of economy, in that up to 1000 doses can be obtained from one chick embryo, rather than the two or three doses possible with a killed vaccine. Live vaccines also induce broader immunity and thus may circumvent the perennial problem of antigenic drift.

Inactivated vaccines are prepared from appropriate strains of influenza A and B grown in the chick embryo allantoic cavity (Chapter 3); the infected fluids are harvested, purified by ultracentrifugation, and inactivated with formalin or beta-propiolactone. Most of those used in Europe, the USA, the CIS (USSR), and Australia are either subunit preparations containing purified HA and NA (Fig. 10.6) or so-called 'split' vaccines that have been extracted with ether and detergent to reduce the side-effects of whole virus vaccines (Table 10.1).

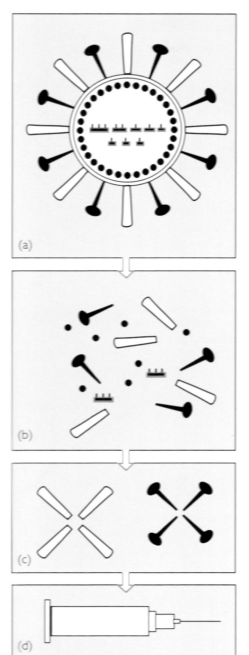

Fig. 10.6. Subunit influenza vaccine. (a) Virus is grown in eggs, purified, and concentrated. (b) Viral lipid is dissolved by detergent, releasing proteins and nucleic acid. (c) HA and NA subunits are purified by ultracentrifugation. (d) Standardized quantities of HA and NA are incorporated into the final product.

Apart from local erythema and soreness, sometimes with fever, these vaccines are generally very safe. There are rare reports of encephalomyelitis and neuritis and, in the mass immunization programme mounted in the USA in 1976 and 1977 to counter the threatened epidemic of swine influenza (see Section 3.1), there was one case of Guillain–Barré syndrome for every 130 000

Table 10.1 Inactivated influenza vaccines

Type	Remarks	Suitable for children
Whole virus	Good immunogen but may cause local and general reactions	No
'Split' vaccine	Whole virus extracted with ether	Yes
'Subunit' or 'surface' antigen	Purified HA and NA extracted with detergent; less liable to cause reactions	Yes

vaccinees. We do not know if this unusual and unpleasant neurological complication was due to the A/swine-like strain of virus used in the vaccine, or whether it came to notice simply because 40 million people were inoculated over a short period.

Live attenuated vaccines are made by hybridizing viruses possessing the required HA and NA antigens with various types of mutant, selected either for inability to grow at 37°C or for adaptation to animals or chick embryos; both types are of diminished virulence for humans. Such live vaccines have been used, often in aerosol form, but, despite their potential advantages, those produced so far seem to be less effective than killed vaccines.

5 REMINDERS

◆ Influenza is caused by the **orthomyxoviruses** which contain **RNA** and have **helical** symmetry; there are three species or types, A, B, and C. Only types A and B cause significant disease in man; they are also prevalent in a number of mammalian and avian hosts and the infection can be transmitted between species.

◆ The **HA** and **NA antigens** are important in the infection of cells, and the corresponding antibodies are important in preventing infection.

◆ The genome has eight segments, allowing **reassortment** of genes; in type A viruses, but not type B, this may give rise to **antigenic shift**, in which the HA, NA, or both differ from those in the previously circulating strains and may derive from animal or avian viruses. Shifted strains appear at long intervals and cause major epidemics.

◆ By contrast, **antigenic drift**, due to minor mutations, affects both A and B viruses, is slowly progressive, and causes more frequent but localized outbreaks.

◆ Clinically, influenza is primarily a short-lived infection of the upper respiratory tract; it is more severe in older people. Complications include **secondary bacterial infection** and, rarely, Reye's syndrome.

◆ The **laboratory diagnosis**, if needed, is made by isolation of the virus in chick embryos or cell cultures or, more rapidly, by **immunofluor-escence staining** of nasopharyngeal washings.

◆ **Amantadine** is moderately effective in prophylaxis and therapy.

◆ The main method of control is at present by means of **inactivated vaccines**, which are, however, usually in short supply because of the constant antigenic changes in the prevalent viruses.

GASTROENTERITIS VIRUSES

1 INTRODUCTION

It is an odd fact that, although the *Enteroviridae* replicate primarily in the gut, they do not, by and large, cause gastroenteritis. Those that do are something of a ragbag which we can term **enteric viruses** to distinguish them from the enteroviruses. It is also curious that most of these agents cannot be propagated in cell cultures, which explains why some of them are poorly characterized.

Table 11.1 lists the viruses associated with diarrhoea and vomiting (D & V). The strength of the association and the relative importance of the various agents are indicated in the last column. We have also given a rough idea of size, which is relevant to their identification: most of them were first recognized by electron microscopy of stool samples, a technique that is still the mainstay of diagnosis. For virologists, some at least of these viruses make up for their inability to grow in the laboratory by their elegant appearance

Table 11.1 Viruses associated with gastroenteritis in humans

Viruses	Size*	Relative importance†
Rotaviruses	L	+ + +
Adenoviruses types 40, 41	L	+ +
'Small round viruses', e.g. Norwalk	S	+ +
Coronaviruses	L	±
Caliciviruses	S	±
Astroviruses	S	±

* L, 'large' (>50 nm in diameter); S, 'small' (<50 nm in diameter).
† Range from + + + (important) to ± (doubtful importance).

under the electron microscope. We shall start off with the rotaviruses since they are the most important and the best understood; the other enteric viruses will be dealt with in less detail.

2 ROTAVIRUSES

2.1 *Properties of the viruses*

Classification

These agents are members of the *Reoviridae*, 'reo' standing for *r*espiratory *e*nteric *o*rphan: although they could be identified in the respiratory tract and gut, they were at first thought not to be associated with any disease, hence 'orphan'. The rotaviruses are a genus within the family; another genus, the orbiviruses, will appear in Chapter 17.

Morphology

The shape of these viruses suggested their name (Latin *rota*: wheel). The capsid has a double shell and is 70 nm in diameter; some smaller, single-shelled particles may also be seen. In most of the virions a number of 'spokes' radiate from a central 'hub', although some appear empty by negative contrast staining. The cores appear to be icosahedral (Fig. 11.1(a)).

Genome

The *Reoviridae* differ from all other RNA viruses in that their genomes are double-stranded. Electrophoresis of extracted RNA shows that they are also divided into 10–12 segments of varying size, coding for a similar number of structural and non-structural proteins. The rotavirus genome contains 11 segments with a total molecular weight of about 11×10^6 (Fig. 11.2). The electrophoretic migration pattern of the nucleic acid differs between strains and is used by some to define 'electropherotypes', which do not correspond with the serotypes but which may be useful for epidemiological studies.

Growth in cell cultures

Unlike other reoviruses, rotaviruses can, if treated with trypsin, be propagated in primary monkey kidney cell cultures, but this method is not used for routine diagnosis.

2.2 Clinical and pathological aspects

Clinical features

■■■■■■ Rotaviruses primarily infect the young of many species, including man. **Babies under 2 years of age** are the main victims but there may also be outbreaks in the **elderly**, particularly those in institutions.

The **incubation period** is **2–4 days**; the characteristic syndrome comprises **vomiting**, **diarrhoea**, **and fever**, but silent infections also occur. **Dehydration** must be dealt with promptly; it should be no problem in countries with adequate facilities, but causes untold numbers of infant deaths in the Third World.

Pathogenesis

■■■■■■ Observations on infected animals show that rotaviruses attack the **columnar epithelium** at the apices of the villi of the **duodenum** and **upper ileum**; the loss of these cells results in malabsorption. Regeneration from the bases of the villi is normally rapid after the acute attack.

Immune response

■■■■■■ Following infection, antibodies are demonstrable in the serum; tests for specific IgG are useful in epidemiological studies, which show that most adults have been infected at some time or another. IgA antibody can be demonstrated in milk, but evidence of its value for protecting suckling infants against infection is conflicting.

Epidemiology

■■■■■■ We tend to associate diarrhoea and vomiting (D & V) with the summer months, but in the Northern hemisphere, rotavirus outbreaks typically occur in the winter. Spread is probably by the **faecal-oral route**; but the seasonal incidence suggests that respiratory infection cannot be ruled out. Outbreaks are common in institutions such as crèches and hospital-acquired infections are also fairly frequent. Unfortunately, the virus is relatively resistant to chemical disinfectants and spreads readily when hygiene is inadequate. In developing countries, infections with rotaviruses, along with other viruses and bacteria, are responsible for large numbers of infant deaths from D & V every year.

Transmission to infants from asymptomatic health care staff or relatives can occur; conversely, there is evidence that babies can infect adults in close contact with them. Chronic rotavirus infection can be troublesome in patients with primary immunodeficiencies (Chapter 26).

The segmentation of the rotavirus genome permits reassortant strains to be made in the laboratory. Since these viruses are widespread in nature and infect most if not all domestic and farm animals and birds, there is at least a theoretical possibility that infection of humans with mammalian or

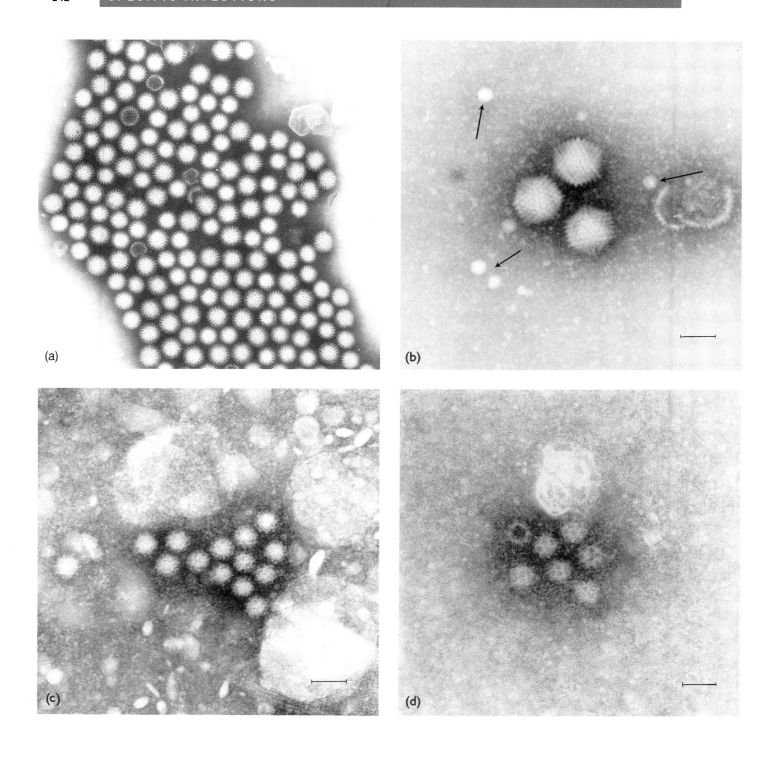

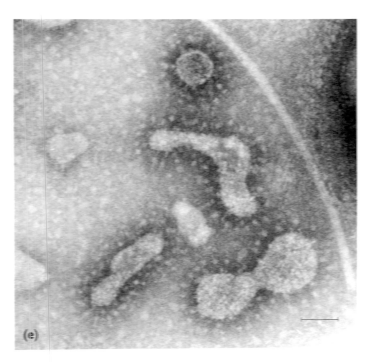

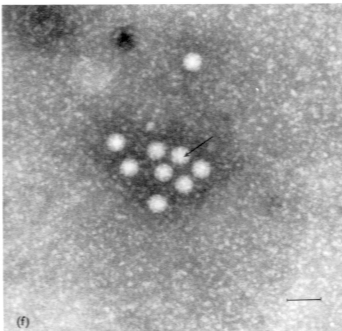

(e)

(f)

avian strains might take place, with the emergence of 'shifted' strains analogous to those of influenza A (Chapter 10), but so far, such hybrids have not been reported.

Rotavirus vaccine

There is no specific therapy for rotavirus infections, so that, especially in view of the problem they pose in developing countries, a vaccine would be a major benefit. During the last few years, promising results have been obtained with vaccines prepared by various methods. They include: (1) the use of bovine strains, which cross-protect to a significant degree against human rotaviruses; and (2) the construction of attenuated reassortant strains, similar in principle to influenza vaccines (Chapter 10).

Fig. 11.1. Electron micrographs of gastroenteritis viruses found in stools of patients with diarrhoea. **(a) Rotavirus**. (Courtesy of Dr June Almeida.) **(b) Adenovirus** from child with diarrhoea. Note also the very small adeno-associated parvoviruses (arrowed). **(c) 'Small round virus'** (SRV). **(d) Caliciviruses** from faeces of child with diarrhoea. Note the cup-like depressions on the surface. **(e) Coronavirus** particles from faeces of HIV-positive patient. Note the characteristic crown-like projections. **(f) Astroviruses**. The six-pointed star pattern of the capsomeres is arrowed. (Parts **(b)-(f)** courtesy of Dr Ian Chrystie.)

Scale bars. 50 nm.

3 ADENOVIRUSES

The relationship of these agents with respiratory and eye infections is well established (Chapter 8); their implication in diarrhoeal disease is often less clear. Like rotaviruses, their distinctive morphology (Fig. 8.1) makes them easy to spot in faecal specimens by electron microscopy (Fig. 11.1(b)), but again, their mere presence does not necessarily indicate that they are doing any harm. There are more than 40 serotypes of adenoviruses, most of which can readily be grown in cell cultures; they can also be arranged by DNA 'fingerprinting' into six subgenera, lettered A to F. Group F contains **serotypes 40** and **41** which differ from other adenoviruses in that they are **'fastidious'**, i.e. difficult to grow in cell cultures. It is these viruses that are most commonly associated with acute diarrhoea in infants; other serotypes, which can readily be propagated from stool cultures, usually seem to be just passengers.

Not uncommonly, adeno-associated parvoviruses can be seen in association with the adenoviruses (Fig. 11.1(b); Chapter 12). They do not seem to be pathogenic.

The clinical syndrome is similar to that caused by rotaviruses, except that the infants tend to be older; adenovirus gastroenteritis is sometimes complicated by intussusception.

4 'SMALL ROUND VIRUSES' (SRV)

The virions (Fig. 11.1(c)) are 25–35 nm in diameter and nondescript in appearance; they cannot as yet be propagated in cell cultures and are poorly characterized. Some have definite morphological features and are termed 'small round structured viruses' (SRSV). The best known is the **Norwalk agent**, first detected in an outbreak of D & V in Norwalk, Ohio. Its association with disease in man has been confirmed by experiments in volunteers. Viruses of this sort are responsible for a substantial number of cases of epidemic gastroenteritis, some of which are associated with infected shellfish.

5 CALICI-, CORONA-, AND ASTROVIRUSES

We have still not exhausted the morphological pleasures provided by electron microscopy of faecal specimens: as well as the elegant rotaviruses, other RNA viruses, equally distinctive — and equally reluctant to grow in cell cultures — are found in association with gastroenteritis, sometimes in circumstances which makes their aetiological role highly probable. The **caliciviruses** (Fig. 11.1(d)) are named for the series of cup-like depressions that characterize their profile; the **coronaviruses** (Fig. 11.1(e)) for their crown-like array of spikes terminating in small spheres; and the **astroviruses** (Fig. 11.1(f)) for the star-shaped pattern of capsomeres that is often visible on their surfaces.

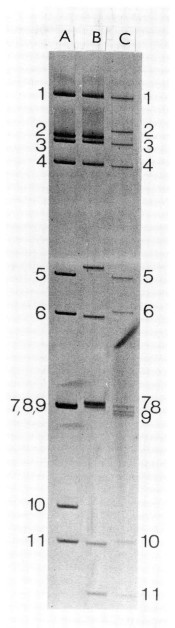

Fig. 11.2. Separation of the gene segments of rotavirus on a polyacrylamide gel. The 11 bands in each channel correspond with the individual genes of three different rotaviruses (A, B, C).

LABORATORY DIAGNOSIS

We have already pointed out that detection of a micro-organism in a clinical specimen does not necessarily mean that it is causally related to the patient's signs and symptoms; enough has been said in this chapter to indicate that this is nowhere more obvious than in acute gastroenteritis. Under the electron microscope, viruses, sometimes of more than one variety, can often be found in faecal specimens. Their association with disease is most probable in the case of rotaviruses but, as far as the others are concerned, epidemiological factors such as the presence of the same virus in the same outbreak must also be considered.

None of this makes the task of the virology laboratory any easier. There are two approaches to diagnosis.

◆ Electron microscopy of faecal extract. This is sometimes called the **catch-all method**, because it is potentially capable of demonstrating all the viruses present in the specimen.

◆ Specific tests aimed at detecting a particular virus.

6.1 *Electron microscopy*

Some viruses commonly seen in stools, e.g. polio, other enteroviruses, and some adenoviruses can also be grown readily in cell cultures and thus identified; these are not associated with gastroenteritis. The viruses we are now hunting are all fastidious, growing poorly or not at all in the laboratory. All the viruses in Table 11.1 are, however, readily identified by their distinctive morphology except for the nondescript 'small round viruses'; for full identification these need sophisticated tests which are not available in the routine diagnostic laboratory.

A useful aid to the finding and identification of viruses in stool samples is **immune electron microscopy**. A specific antiserum added before the specimen is mounted and stained will clump any virions of the same specificity, at one stroke making them easier to find and proving their identity. The method can be used for rota-, adeno-, and enteroviruses.

6.2 *Specific tests*

Laboratories without electron microscopes are at a disadvantage, but can fall back on other methods, of which the most popular are ELISA tests (Chapter 27) applied to stool samples; at present, only tests for rota- and adenoviruses are generally available. Specific antibodies can also be used to identify some fastidious viruses that undergo limited replication in cell cultures without inducing cytopathic effects.

Polyacrylamide gel electrophoresis (PAGE) can be used to identify the nucleic acids of specific viruses such as rota- and adenoviruses but this method, although promising and technically simple, is not yet commercially available.

It should by now be apparent that attaching the right label to single cases of acute gastroenteritis is fraught with difficulties and that the best chance of success lies in the study of outbreaks in which the presence of a particular virus is a common factor. Such investigations demand the closest co-operation between physicians and laboratory staff; the haphazard collection and examination of specimens is useless.

7 REMINDERS

◆ Acute gastroenteritis of viral origin is associated not with the *Enteroviridae* but with a number of miscellaneous agents including rotaviruses, certain adenoviruses, 'small round viruses', and, less definitely, with calici-, corona-, and astroviruses.

◆ These 'enteric viruses' are '**fastidious**', i.e. they grow poorly or not at all in cell cultures.

◆ The most important are the **rotaviruses**, which, like the other *Reoviridae*, are unique in having a double-stranded segmented RNA genome. Spread is primarily by the **faecal–oral** route. They cause outbreaks of gastroenteritis in young babies and sometimes in old people. In temperate countries, rotavirus infections occur in **winter** rather than in summer and are particularly liable to affect those in institutions. Rota- and other 'enteric' viruses are a major cause of infant mortality in Third World countries.

◆ Diagnosis of enteric virus infections depends mainly on **electron microscopy** of stool specimens. **ELISA** tests for rota- and adenoviruses are also routinely available in laboratories. It is emphasized that the mere finding of a particular virus in a stool specimen does not prove its relationship with the disease; other factors, particularly epidemiological, must also be considered.

◆ Trials of **rotavirus vaccines** are yielding promising results.

PARVOVIRUSES

1 INTRODUCTION

It might be thought that viruses as a class represent the ultimate in parasitism, reliant as they are on their host cells to provide most of the machinery for replication. The parvoviruses, however, show a still further degree of dependence, since they can replicate only in the presence of another virus or of active DNA synthesis in rapidly dividing host cells. The reason lies in their minute size, for these are the smallest of all viruses (Latin *parvus*: small).

2 PROPERTIES OF THE VIRUSES

2.1 *Classification*

There is one family, the *Parvoviridae*, containing three genera, two of which infect humans (Table 12.1). The **dependoviruses** are named for their dependence on a **helper virus**, usually an adeno- but occasionally a herpesvirus, to assist in replication. The **Parvovirus** genus takes its name from that of the family. Several strains have been described: B19 is most often implicated in infections of man. Agents resembling parvoviruses have been associated with gastroenteritis acquired from shellfish. Other parvoviruses infect rodents, cats, dogs, and other animals. The genus *Densovirus* infects only invertebrates.

Table 12.1 Human parvoviruses

Genus	Viruses	Diseases
Dependovirus	Adeno-associated viruses (5 types)	Not pathogenic
Parvovirus	B 19	*Erythema infectiosum*; aplastic crises
	RA–1	Possible implication in rheumatoid arthritis

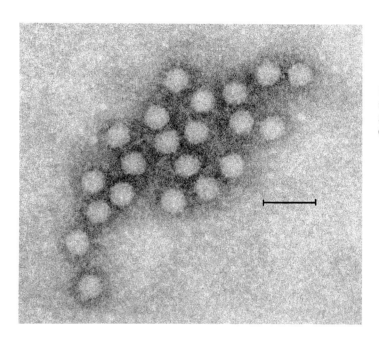

Fig. 12.1. Electron micrograph of parvovirus in serum. The virions are clumped together by antibody molecules. (Courtesy of Dr Ian Chrystie.) Scale bar, 50 nm.

2.2 *Morphology*

These are non-enveloped viruses, with icosahedral symmetry, about 22 nm in diameter (Fig. 12.1).

2.3 *Genome*

The *Parvoviridae* are the only DNA viruses with **single-stranded** (ss) nucleic acid. They were thought to be double-stranded until it was discovered that each virion contains either a positive- or a negative-sense strand, a unique situation. A 'hairpin' loop at the end of each genome initiates replication of the complementary strand. The ssDNA is 5.5 kb long.

2.4 *Polypeptides*

Such a small genome can code only for a limited number of polypeptides. There are only three, of which two are structural; this explains why these agents are so dependent on external helper functions.

2.5 *Replication*

The dependoviruses used to be called 'adeno-associated viruses' since they are seen by electron microscopy in association with enteric adenoviruses (Chapter 11). The genome becomes integrated into host-cell DNA and is active only when the cell becomes infected with a helper virus that can provide the enzymes essential for replication. The parvoviruses, on the other hand, replicate only in rapidly dividing host cells, in which S-phase functions provide the necessary help. No parvoviruses have yet been grown *in vitro*.

 # 3 CLINICAL ASPECTS

The dependoviruses are not pathogenic. The parvoviruses infecting humans cause two main types of illness.

3.1 *Erythema infectiosum*

This 'infectious rash' is sometimes known as 'fifth disease'. (The original classification of the exanthematous illnesses of childhood comprised six diseases: the other five are measles, rubella, scarlet fever, exanthem subitum, and Duke's disease, a rash of obscure aetiology.) A more picturesque name is 'slapped cheek syndrome', which helps you to remember that the presenting feature is often a **maculopapular rash over the malar areas**, followed within the next 4 days by a rash on the trunk and limbs which may persist for 2 or 3 weeks. The incubation period is 13–18 days. There may be some fever and malaise in the early stages, and mild febrile illness without rash is common.

This is predominantly an infection of children but adults may also acquire it; in them, and **particularly in women**, the **joints** are much more likely to be involved. Those of the hand and fingers are most often affected, but there may also be arthropathy of the arms, legs, and spine. Arthralgia may persist for a few weeks but there appear to be no long-term sequelae. Infection in **pregnancy** may result in fetal death (Chapter 25).

3.2 *Aplastic crisis*

The need of parvoviruses for actively dividing cells in which to replicate is illustrated by their lytic effect on red-cell precursors: this is not clinically obvious in otherwise healthy people, but in those with a chronic haemolytic anaemia, e.g. sickle cell anaemia or thalassaemia, parvovirus infection may precipitate an **aplastic crisis** with very low haemoglobin and disappearance of circulating reticulocytes. This is because the virus replicates in and damages the rapidly dividing late normoblasts.

Persistent infection with B19, associated with chronic anaemia, has been observed in a variety of conditions associated with immunodeficiency, including acute lymphocytic leukaemia and HIV infection.

4 PATHOGENESIS AND IMMUNE RESPONSE

Volunteer and field studies show the incubation period to be 13–18 days. After primary infection of the **upper respiratory tract** there is a **viraemia** lasting about a week. Clearance of virus from the blood coincides with a sharp IgM antibody response, followed shortly by the appearance of IgG antibody. The rash and arthropathy in erythema infectiosum are probably immunologically mediated. Reference has already been made to the role of cell lysis in haemolytic crises.

5 EPIDEMIOLOGY

Spread is by the **respiratory route**: occasional transmissions by blood transfusion have been reported. Epidemics occur during **late winter** and **early spring** and mainly affect **young schoolchildren**.

6 DIAGNOSIS

6.1 *Clinical diagnosis*

Erythema infectiosum may be diagnosed by the characteristic rash, especially during an epidemic, but differential diagnosis from other illnesses with fever and rashes is not always easy. The syndrome of fever, rash, and arthropathy may be clinically indistinguishable from rubella but it is essential to establish the correct diagnosis with absolute certainty in pregnancy; fortunately, laboratory tests can give the answer.

6.2 *Laboratory tests*

These viruses cannot be grown in the laboratory, but can be found in the blood by **electron microscopy** during the viraemic stage. Virus DNA can also be identified by dot hybridization. **Detection of IgM antibody by ELISA** or **radioimmunoassay (RIA)** indicates a current or recent infection.

7 REMINDERS

◆ The *Parvoviridae* are the smallest viruses and the only ones containing **ssDNA**. Virions contain either positive- or negative-sense strands.

◆ The dependoviruses need help from adeno- or herpesviruses in order to replicate. They are not pathogenic.

◆ The parvoviruses cause:

1. **Erythema infectiosum**, a febrile illness with rash that is spread by the respiratory route, mainly in young children. Cure is spontaneous and there are no known sequelae. In adults, especially women, there may be arthropathy.

2. **Aplastic crises** in people with pre-existing chronic haemolytic anaemia, e.g. thalassaemia or sickle cell disease.

◆ The differential diagnosis of erythema infectiosum may be difficult clinically, but can be made by appropriate laboratory tests. In pregnancy, it is particularly important to distinguish it from rubella.

POXVIRUSES

1 INTRODUCTION

In a book of this size there is a great temptation to omit material that is no longer directly relevant to current practice; but it would be unthinkable to discuss the poxviruses without an account, albeit brief, of smallpox, the great success story in the fight against infectious disease and one that provides many valuable lessons. This epic provides at least three 'firsts': the first vaccine; the first disease to be totally eradicated by immunization; and the first virus infection against which chemotherapy was clinically effective. Although smallpox itself is now extinct, other poxviruses infect humans and, as we shall see in Chapter 28, one of them may be used in a rather surprising way to immunize against quite different diseases.

2 PROPERTIES OF THE VIRUSES

2.1 *Classification*

The Family *Poxviridae* contains eight genera. Most of these viruses affect only species other than man, ranging literally from insects to elephants. Others are pathogenic both for animals and humans and two, smallpox and molluscum contagiosum, only for man (Table 13.1). The genera are distinguished on the basis of morphology, genome structure, growth characteristics, and serological

Table 13.1 Poxviruses that infect humans

Genus	Virus	Primary host(s)	Clinical features in humans
Orthopoxvirus	Variola	Man	Smallpox
	Vaccinia	Man	Vesicular vaccination lesion
	Cowpox	Cattle, cats, rodents	Lesions on hands
	Monkeypox	Monkeys, squirrels	Resembles smallpox
Parapoxvirus	Pseudocowpox	Cattle	Localized nodular lesions ('milkers' nodes')
	Orf	Sheep, goats	Localized vesiculogranulomatous lesions
Unclassified	Tanapox	Monkeys	Vesicular skin lesions and febrile illness
	Molluscum	Man	Multiple small skin nodules

reactions; there is close serological relationship between the viruses within each genus and good cross-protection between genera.

2.2 *Morphology*

These are the largest viruses of all; the orthopoxviruses are brick-shaped, whereas orf and molluscum contagiosum tend to be ovoid (Fig. 13.1). They measure about 230×270 nm and, when suitably stained, can just be seen with an ordinary light microscope. The poxviruses are neither icosahedral nor helical: their structure is referred to as **complex**. The capsid consists of a network of tubules and is sometimes surrounded by an envelope.

2.3 *Genome and polypeptides*

The nucleic acid is **dsDNA** with a molecular weight of around 10^8. This large genome codes for more than 100 polypeptides, including a DNA-dependent RNA polymerase and other enzymes. A number of these polypeptides are antigenic and are thus to some extent useful in classification and diagnosis.

2.4 *Replication*

Unlike most DNA viruses, the **poxviruses replicate only in the cytoplasm**, in which they form inclusion bodies. Both cell and virus-coded enzymes are involved in uncoating. Viral DNA-dependent RNA polymerase then mediates

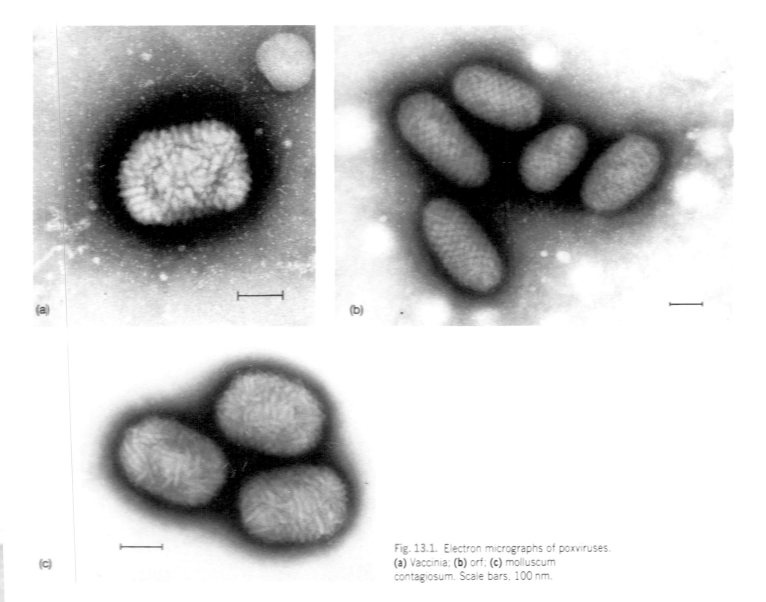

Fig. 13.1. Electron micrographs of poxviruses.
(a) Vaccinia; (b) orf; (c) molluscum
contagiosum. Scale bars, 100 nm.

transcription of mRNA from the input genome; the general pattern of replication follows that for DNA viruses (Chapter 2).

2.5 *Propagation in the laboratory*

Most poxviruses can be propagated on the chick embryo chorioallantoic membrane, on which they form circumscribed pocks, 2–3 mm in diameter, or in cell cultures. Molluscum contagiosum has not so far been grown in the laboratory.

3   CLINICAL AND PATHOLOGICAL ASPECTS

3.1   *Smallpox*

History

The Anglo-Saxon word *pokkes* meant a pouch and refers to the characteristic vesicular lesions. The term 'smallpox' was introduced during the sixteenth century to distinguish it from the 'great pox', or syphilis. The Latin term, *variola*, means a spot. It seems likely that smallpox has been with us for a very long time: the mummy of the Pharaoh Rameses V (1100 BC) bears lesions highly suggestive of this infection (Fig. 13.2) and there are many later accounts of its ravages. Indeed, smallpox was so widespread that it was often regarded as the norm rather than the exception and few people — if they survived — escaped its disfiguring scars. In India it was thought to be a divine visitation and even had its own goddess, Kakurani.

The observation that smallpox could be prevented by inoculation of healthy people with material from the lesions seems to have originated in China. From there an account of the practice (**variolation**) was sent by Joseph Lister to The Royal Society in 1700 and was followed by others from Turkey.

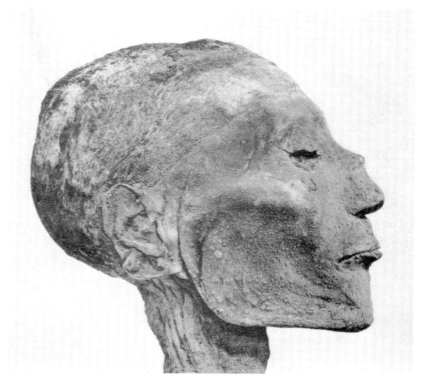

Fig. 13.2. Mummy of Rameses V. The lesions on the face are suggestive of smallpox. (Reproduced, with permission, from Dixon, C.W. (1962). *Smallpox*. J&A Churchill.)

Fig. 13.3. Lady Mary Wortley Montague. (Courtesy of the Wellcome Institute Library, London.)

Variolation caused a comparatively mild infection from which the subject usually recovered and which effectively protected against the natural disease.

There now enters one of the most interesting characters in the story, Lady Mary Wortley Montague (Fig. 13.3). As wife of the British ambassador, this highly accomplished young woman saw variolation practised in Turkey and was so convinced of its safety and efficacy that in 1717 she had her own small son inoculated. Lady Mary had a vested interest in the subject because she herself had been scarred by the disease; on returning to England she used all her efforts and her many highly-placed contacts to have the practice generally accepted. She clearly anticipated opposition from the medical establishment for she wrote 'I am patriot enough to take pains to bring this useful invention into fashion in England, and I should not fail to write to some of our doctors very particularly about it, if I knew any one of them that I thought had virtue enough to destroy such a considerable branch of their revenue, for the good of mankind.' In the event, variolation enjoyed but a short vogue before being superseded by the safer method of **vaccination**.

The story of Edward Jenner's discovery in 1796 that inoculation with cowpox would prevent smallpox is well known; but it is not so widely appreciated that others had made similar observations, notably Benjamin Jesty, a Dorset farmer, who inoculated his own family. To Jenner, however, goes the credit for showing that, following the inoculation of young James Phipps with cowpox, deliberate inoculation with smallpox material failed to induce the disease (Figs. 13.4 and 13.5).

The early vaccines were derived from cowpox (Latin *vacca* = cow) and propagated by arm-to-arm inoculation. During the last century other strains of poxvirus (vaccinia) that could be grown in quantity on the skins of calves and other animals were developed.

With improved methods for preparing vaccine in bulk and for testing its safety and potency, vaccination was widely practised; nevertheless, wide regional differences in uptake ensured that the disease would continue to smoulder on, flaring at intervals into epidemics. In 1966, faced with this situation, the World Health Organization (WHO) voted US$2.5 million for an

Fig. 13.4. Edward Jenner and James Phipps. (Courtesy of the Wellcome Institute Library, London.)

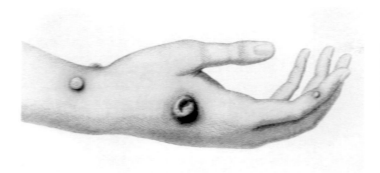

Fig. 13.5. Cowpox lesions on the arm of the milkmaid, Sarah Nelmes. Material from these lesions was used by Edward Jenner to vaccinate James Phipps. (Courtesy of the Wellcome Institute Library, London.)

immunization campaign designed to eradicate smallpox completely within the next decade. The saga of this vast enterprise is referred to again in Chapter 28, but needs a book to itself. The full story is told in *Smallpox and its eradication*, by F. Fenner *et al.*, published in 1988 by the World Health Organization. Its success may be judged by the fact that the last case of naturally acquired infection was recorded — in Somalia — in October 1977, only 10 months beyond the target date fixed 10 years previously (Fig. 13.6).

Clinical aspects

It is very gratifying to be able to write this section in the past tense!

There were two main categories of smallpox, caused by slightly different viruses: **variola major** had a mortality of around 30 per cent whereas **variola minor**, or **alastrim**, killed less than 1 per cent of its victims. The **incubation period** was usually 10–12 days, with a range of 8–17 days; a febrile illness of sudden onset lasting 3–4 days was followed by the appearance of a rash progressing from macules to papules, vesicles, and pustules which then formed crusts (Fig. 13.7). Surviving patients were often left with unsightly scars or pockmarks. The distribution of the rash was **centrifugal**, i.e. it affected the extremities more than the trunk, as opposed to the centripetal rash of chickenpox. Haemorrhagic and fulminating forms occurred which were rapidly lethal. Modified smallpox with few lesions and comparatively little constitutional upset was sometimes seen in people who had been vaccinated some years previously.

Pathogenesis

The pattern was that of an acute generalized infection (see Chapter 4, Section 4.4). The lesions were caused by direct viral invasion and the vesicles contained large numbers of virions.

Immune response

An attack nearly always conferred life-long immunity, mediated by neutralizing antibody. Recovery from infection was, however, largely effected by cell-mediated responses.

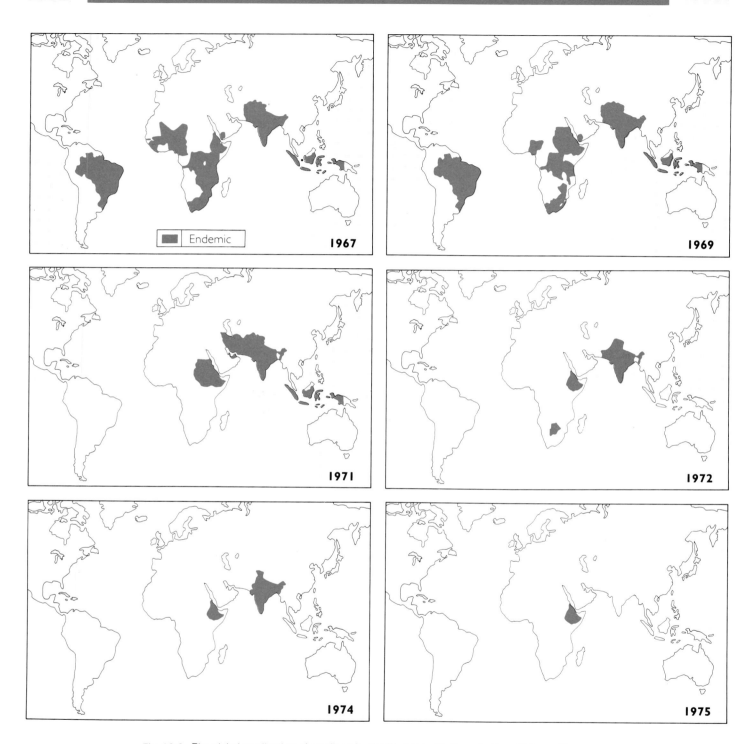

Fig. 13.6. The global eradication of smallpox by the World Health Organization. (Taken with permission from Fenner. F. et al. (1988). Smallpox and its eradication, World Health Organization, Geneva.)

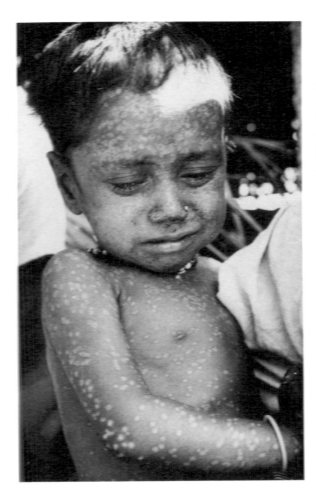

Fig. 13.7. Smallpox. This 2-year old Bangladeshi girl was the last case of smallpox to be seen on the Asian subcontinent, and the last case in the world of variola major, the more severe form of the disease. (Photograph from World Health Organization.)

Epidemiology

The disease was mostly prevalent in the Eastern countries, the Indian subcontinent, and Latin America, both in endemic and epidemic form. Spread from country to country was facilitated by increases in the volume and speed of international travel, and, in the absence of universal vaccination, importations into Europe and North America occurred with some frequency well into this century. 1962 saw the last outbreak in the United Kingdom; the last case, however, was a fatal laboratory-acquired infection in 1978.

Transmission. The main source of infection was the patient's **upper respiratory tract** in the early stages of the disease, but fomites such as bedding and clothing were also of some importance.

Laboratory diagnosis

The differential diagnosis most often needed was between smallpox and chickenpox, which could on occasion resemble each other clinically. The introduction of electron microscopy, which readily and rapidly distinguished between pox and herpes virions in vesicle fluid, was most valuable in this respect.

Prevention and treatment

Prophylaxis depended on vaccination, which not only induced immunity to subsequent exposure, but also protected if given early enough during the 12-day incubation period. Epidemics were contained by mass vaccination until the later stages of the world eradication programme, when this approach was replaced by selective vaccination of contacts.

There was no specific treatment. **Methisazone**, a thiosemicarbazone, would prevent or modify an attack if given during the incubation period but, soon after the discovery of this, the first effective antiviral compound, it was made redundant by the success of the vaccination campaign.

Survival of variola virus

Following tragic episodes of laboratory-acquired infection it was internationally agreed that samples of the virus should be held only in two maximum security laboratories, one in Atlanta, USA and the other in Moscow; furthermore, large stocks of vaccine are held in store for use in the unlikely event of a reappearance of the disease. Even so, in view of the depredations of smallpox and the herculean efforts needed to eradicate it from the planet, the Committee on Orthopoxvirus Infections recommended that all remaining stocks of variola virus should be destroyed. The complete destruction of this virus is a controversial issue, but in our view would be a mistake; however nasty, every living species embodies a unique fund of biological information and the obliteration of any one of them represents an irreplaceable loss to future generations. A compromise solution to the dilemma was found by cloning the viral DNA into bacterial plasmids in such a way that it cannot be expressed, but still embodies the genetic information as a permanent record of one of the great viral plagues of mankind.

In view of the ability of poxviruses to survive for long periods in the dried state, it has been suggested that infective smallpox virus might still exist in preserved cadavers; this seems unlikely, but it is interesting that in 1986, Italian workers claimed to have demonstrated smallpox virus in a Neapolitan mummy dating from the sixteenth century AD.

3.2 *Vaccinia and smallpox vaccine*

We mentioned in Section 3.1 that strains of poxviruses other than cowpox were developed for making smallpox vaccines. The origins of many of these strains, which became known as **vaccinia**, are obscure; some were derived from cowpox, whereas others may have been attenuated strains of smallpox. The virus was propagated in large quantities on the skins of calves, sheep, or — in the East — water-buffaloes. The crude material was treated to reduce the content of skin bacteria, and kept in a glycerol solution at subzero temperatures until used. Such vaccines were, however, unstable at ambient temperatures, particularly in the tropics, and the success of the World Health Organization eradication campaign depended largely on the heat-stable freeze-dried vaccine developed at the Lister Institute of Preventive Medicine, Elstree, UK.

Inoculation of a small quantity of vaccinia into the superficial layers of the skin resulted in a vesicular lesion that was well developed 7–9 days later and rapidly evoked cross-immunity to variola; thus vaccination was effective if performed during the early part of the 12-day incubation period of smallpox.

3.3 Cowpox

Like most of the agents in this family, cowpox is, as its name implies, a zoonosis (Table 13.1); infection of humans is rare but may be seen in rural practices. Although recent evidence points to cats and rodents rather than bovines as the main reservoir of infection, most cases in humans seem to be acquired from cows with sores on their teats. The lesions usually present on the hands or face and resemble severe vaccinia infections.

Treatment with antivaccinia immunoglobulin may be helpful if given early.

3.4 Monkeypox

As far as is known, this zoonosis occurs only in the rain-forest areas of West Africa; it is of some concern because it causes an illness in humans very similar to smallpox, with a percentage mortality running well into double figures. The infection is seen mainly in children who may acquire it from playing with captive animals. So far, there have been more than 300 cases but, since human-to-human infection is infrequent, control by vaccination is not difficult.

Could monkeypox mutate to smallpox? The question was raised as soon as infections in humans were recognized; but fortunately the poxviruses, like others with DNA genomes, are genetically very stable, so that this dire possibility can, one hopes, be ignored.

3.5 Parapoxviruses

Pseudocowpox infects various species of cattle whereas **orf** is found in sheep and goats in which it causes contagious pustular dermatitis; the viruses are very similar. In humans, lesions occur on the hands and face after contact with infected farm animals; the lesions of pseudocowpox are nodular ('milk-ers's nodes'), whereas those of orf are granulomatous. Parapoxvirus lesions are characteristically painless and resolve over a period of weeks without specific treatment.

3.6   *Tanapox*

This infection takes its name from the Tana River in Kenya, where it was first diagnosed. It is prevalent in monkeys in Kenya and Zaïre but, unlike mon-keypox, appears to be spread by insect bites. There is usually only one vesicular lesion but its appearance is preceded by fever and quite severe malaise. Recovery is uneventful.

3.7   *Molluscum contagiosum*

Like tanapox, this virus is not as yet assigned to a genus, and its study is not helped by our inability to grow it in the laboratory. **Molluscum contagiosum** affects only humans; it is a comparatively common skin condition, character-ized by multiple small nodular lesions, mostly on the trunk (Fig. 13.8). They become umbilicated and contain caseous material in which 'molluscum bodies' can be readily demonstrated. These are quite large (30 μm long) ovoid struc-tures containing many virions (Fig. 13.9). The diagnosis is usually obvious from the clinical appearance, but, in case of doubt, a simple test is to identify

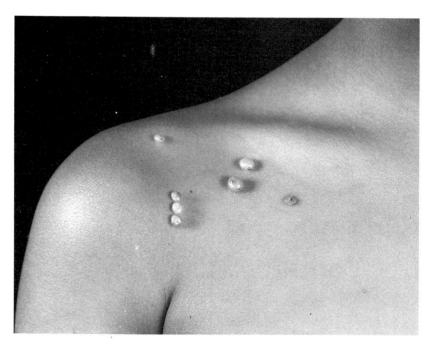

Fig. 13.8. Molluscum contagiosum. (Courtesy of the Department of Dermatology, The Royal London Hospital, London.)

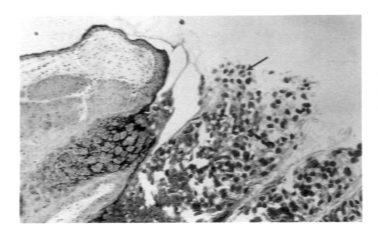

Fig. 13.9. Section of a molluscum contagiosum lesion of the skin. The small ovoid molluscum bodies (arrowed) are packets of virus particles.

the molluscum bodies in expressed material stained with Giemsa or Lugol's iodine. The virions themselves can readily be found by electron microscopy.

Transmission is by contamination of skin abrasions through contact; the infection can be sexually acquired. A molluscum contagiosum lesion at the lid margin often results in severe conjunctivitis and keratitis, which resolves when the nodule is treated.

Disappearance of the lesions is often spontaneous, but can be helped along by cryotherapy, curettage, or treatment with caustic agents such as phenol.

3.8   *Ectromelia*

Ectromelia, or mousepox, does not infect man, but is mentioned here for two reasons. First, it provided the model for Dr Frank Fenner's research on the pathogenesis of acute viral infections in man (Chapter 4), and, second, it may become endemic — and very difficult to eradicate — in stocks of laboratory mice.

4   REMINDERS

◆   The *Poxviridae* are comparatively large, brick-shaped or ovoid viruses with a **complex** structure. Their genome is **dsDNA**. They replicate only in the cytoplasm.

◆   Most poxviruses cause zoonoses, but some are pathogenic for man. Of these, smallpox was the most important until eradicated in 1977 by the WHO vaccination campaign. It was an acute generalized infection: the mortality of variola major was about 30 per cent and that of a milder form, variola minor or alastrim, less than 1 per cent.

◆ **Monkeypox** is a similar infection and occurs in West Africa; it is occasionally transmitted to humans, but person-to-person spread is rare.

◆ **Cowpox, pseudocowpox,** and **orf** affect farm animals and occasionally cause local lesions, which resolve spontaneously, on the fingers and faces of people coming in contact with them. Cowpox may be treated with antivaccinia serum.

◆ **Molluscum contagiosum** causes multiple small nodular lesions containing packets of virions known as molluscum bodies. A lesion at the lid margin may cause **conjunctivitis** and **keratitis**. The lesions may eventually disappear spontaneously, but may be treated by cryotherapy, curettage, or caustic chemicals.

CHAPTER FOURTEEN

POLIOMYELITIS AND OTHER ENTEROVIRUS INFECTIONS

1 INTRODUCTION

The family *Picornaviridae* is one of the largest to be considered in this book and contains some of the smallest viruses, of which poliovirus is the most important. Poliomyelitis was one of the first diseases to be recorded; an Egyptian tomb carving of the Nineteenth Dynasty shows the dead man to have had a foot-drop deformity typical of paralytic poliomyelitis (Fig. 14.1).

The extraordinarily wide range of diseases caused by different members of the family is summarized in Table 14.1. The syndromes include asymptomatic infection, which, fortunately, is by far the most common, disease of the central nervous system (CNS), febrile illness with rash, conjunctivitis, herpangina, infections involving the muscles and heart, and hepatitis. Probably no other family of viruses causes such a diversity of illnesses.

2 PROPERTIES OF THE VIRUSES

2.1 Classification

Enteroviruses are found in several mammalian species, but we shall describe only those that infect humans. These agents are spread by the faecal–oral route and are acid-resistant, which enables them to survive passage through the stomach. The name of the genus derives from the ability of the viruses to

Fig. 14.1. Egyptian tomb carving of the nineteenth dynasty. The 'foot-drop' deformity is characteristic of residual paralysis due to poliomyelitis.

replicate in the small intestine; they can be readily isolated from faeces. Gastroenteritis, however, is not a major feature of enterovirus infections.

Poliovirus was the first of the genus—indeed, of the whole family— to be isolated. This accomplishment by Enders, Weller, and Robins in 1948 earned them a Nobel Prize, not just because poliomyelitis was then highly prevalent in the USA and elsewhere, but because it was the first occasion on which a neurotropic virus was grown in non-neural tissue, in this instance derived from a human embryo. There are three serotypes, originally called Brunhilde, Lansing, and Leon, but now known as types 1, 2, and 3, which are readily distinguished in the laboratory by neutralization tests.

Soon after the isolation of the polioviruses, similar viruses were isolated that paralysed infant mice. On the basis of the pathological changes in mice, two groups were defined, termed coxsackieviruses A and B after Coxsackie, New York, where they were first isolated. (Note that by convention we write 'coxsackievirus A', but 'Coxsackie A virus')

In the late 1950s, with the increasing use of mammalian tissue cultures, yet more viruses of the same sort were isolated from faecal samples, which, however, did not at that time appear to be pathogenic for animals or humans. Such viruses are called **echoviruses**, a term derived from *enteric*

Table 14.1 The *Picornaviridae*

Genus	Main syndromes
Enterovirus	Infections of the central nervous system, heart, skeletal muscles, skin, and mucous membranes; hepatitis A
Rhinovirus	Common colds
Aphthovirus	Foot and mouth disease (rarely in humans)
Cardiovirus	Encephalitis and myocarditis in rodents only

Table 14.2 Enteroviruses that infect humans

Group	No. of serotypes
Poliovirus	3
Coxsackie A	23
Coxsackie B	6
Echovirus	31
Enterovirus	5*

* Numbered 68–72. Type 72 is hepatitis A virus (Chapter 20).

cytopathogenic *h*uman *o*rphan viruses — orphan because they seemed to be viruses that lacked a disease.

So many enteroviruses are now known that those newly identified are referred to by number only (Table 14.2). Enterovirus 72, the cause of hepatitis A, is dealt with in Chapter 20.

2.2 Morphology

Picornaviruses are only 18–30 nm in diameter and **icosahedral**, with a regular protein capsid composed of 32 capsomeres containing four structural proteins; there is no envelope (Fig. 14.2).

Some remarkable studies have recently been performed with polio- and rhinovirus using X-ray crystallography. These viruses have many similarities. But, most important, these analyses have helped to pinpoint functional areas in the virion, the most interesting being the 'canyon' or cleft where the cell receptor-binding site is located, and the immunogenic sites on the exposed external parts of the capsid. These antigenic areas are encoded by hypervariable regions in the rhinovirus and poliovirus genomes; they are important binding sites for neutralizing antibody and hence important in the immunogenicity of poliomyelitis vaccines.

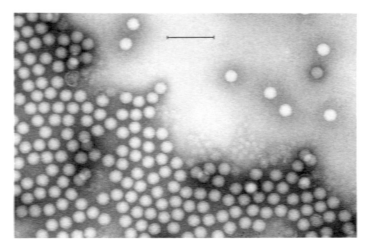

Fig. 14.2. Electron micrograph of a picornavirus (poliovirus). (Courtesy of Dr David Hockley.) Scale bar, 100 nm.

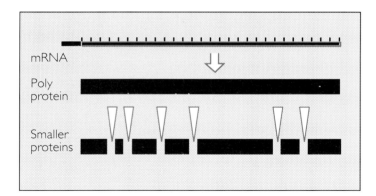

Fig. 14.3. Proteolytic cleavage of poliovirus polyprotein to yield structural proteins. The viral mRNA encodes a single large polyprotein which is then cut by proteases at precise positions ($\triangledown$) to produce a number of smaller proteins with different functions.

2.3 *Genome and polypeptides*

Picornaviruses are **positive-strand RNA** viruses and the genome of some 7500 nucleotides is thus infective for cells.

The virion RNA, which acts as the viral messenger, has a single large open reading frame for translation into one very large polyprotein with a molecular weight approaching 240 000. The four viral capsid proteins (VP1, VP2, VP3, and VP4) and about 20 other proteins are derived from the polyprotein by a series of proteolytic cleavages (Fig. 14.3).

2.4 *Replication*

Replication of viral RNA starts within 1 hour of infection of a cell and has two stages. The parental positive-sense RNA strand is transcribed into a negative-sense strand which serves over and over again as a template for transcription into new progeny positive-sense RNA strands. Assembly of nucleic acids and structural proteins takes place in the host-cell cytoplasm.

3 CLINICAL AND PATHOLOGICAL ASPECTS

3.1 *Poliomyelitis*

This term is derived from the Greek *polios*: grey, and *muelos*: marrow. The shortened form 'polio' is often used to indicate the paralytic form of the disease.

Clinical features

The incubation period is usually 7–14 days, with extremes of 2–35 days. The course is variable (Fig. 14.4); there may be inapparent infection; a minor illness with malaise, fever, and sore throat; or a major illness heralded by a meningitic phase. Each of these, except the major illness, may resolve without sequelae.

The minor illness is due to viraemia. As well as the non-specific signs mentioned, there may be a personality change, especially in young children. In those progressing to the major illness there maybe a few days of apparent well-being before the onset of the meningitic phase, giving a biphasic presentation.

The major illness may occur with or without preceding symptoms; the onset is abrupt, with headache, fever, vomiting, and neck stiffness. The meningitic phase often concludes in a week. However, in a minority of persons paralysis sets in. Its extent ranges from part of a single muscle to virtually every skeletal muscle, in which case severe impairment of respiration may demand the use of artificial ventilation. Paralysis is of the **lower motor neuron type** with **flaccidity** of affected muscles. In **bulbar poliomyelitis** involvement of cranial nerves results in paralysis of the pharynx, again bringing difficulty with respiration.

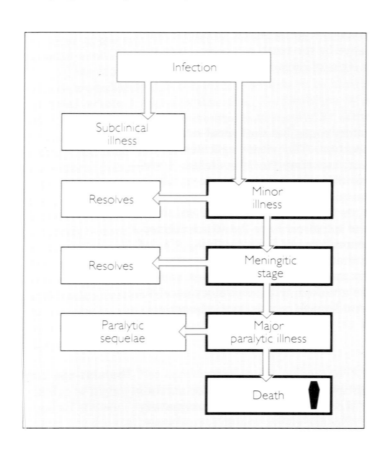

Fig. 14.4. Possible outcomes of infections with poliovirus. Most infections are asymptomatic; only a minority proceed to paralysis.

Possible mutation to neuro... | | very rare

Contra-indicated in immunodeficiency states or pregnancy | No | Yes

'enteroviruses'. The subgroup numbered 68–71 will... referred to by their type numbers.

Table 14.4 lists the diseases caused by these agents; it is apparent that there is much overlap between the syndromes caused by the various groups

however, **lameness surveys** initiated by the World Health Organization (see 'Future control of polio in developed and developing countries', below) have revealed gross under-reporting of paralytic poliomyelitis, particularly in young children, sometimes by factors of 10- or 20-fold.

the poliovaccine strains in the gut (in the same way that one serotype in OPV interferes with the replication of others). Recently, however, a more prosaic factor was discovered, which may account at least for a proportion of vaccine failures: it is simply that in some areas insufficient attention was paid to

3.6 *Laboratory diagnosis of enterovirus infections*

Warning! Because these viruses are so widespread, particularly in children, it is essential to bear in mind that isolation of a particular agent does not necessarily prove its causal relationship with the patient's illness; all the findings—clinical, virological, and serological—must be taken into account. The exception is when virus is isolated from the CSF; this is strong evidence of association with a CNS infection.

Specimens. As a routine, throat swabs, faeces, and serum are collected during the acute phase. Since excretion of virus may be intermittent, it is prudent to take further samples of faeces if the first is negative. Samples should also be taken from vesicular lesions, if present, since the fluid contains virus.

Virus isolation

Human or monkey cells must be used since only primate cells possess the specific receptors for human enteroviruses. All these viruses cause a somewhat similar cytopathic effect within 2–3 days of inoculation; it consists of rounding up of cells and eventually, complete destruction of the monolayer. It is necessary to identify the virus concerned by neutralization tests. In view of the many viruses involved, this is best done by using several pools of reference sera, each containing antibodies to some viruses but not others. By noting which pools do or do not neutralize infectivity, it is possible to identify the virus in the specimen. These reagents are known as LBM pools (after Lim, Benyesh, and Melnick, who devised them). Such tests can be done only in specialist laboratories.

If a coxsackievirus infection is suspected, clinical specimens or infected cell cultures are inoculated into **newborn mice**, in which Coxsackie A and B viruses induce flaccid and spastic paralysis respectively. This is the only virus diagnostic test still employing animals; it will certainly be superseded by tests **in vitro**.

Detection of antibody

A rising titre of antibody in paired sera taken early in infection and 10–14 days later is good evidence of infection, but such tests are of little value in practice, both because of the time involved and the many possible antibodies. They are now being superseded by IgM antibody capture ELISA tests (see Chapter 27); these again are limited by the number of viruses and their corresponding antibodies, but may be useful when infection with a specific virus, e.g. a Coxsackie B, is suspected.

Detection of viral antigens

There are already a number of publications describing the identification by hybridization or by polymerase chain reaction (PCR) of specific viral RNA sequences in tissues infected with enteroviruses; such techniques are not yet available for routine tests, but will undoubtedly play a major part in research on these infections, and perhaps in their diagnosis.

4 CONTROL MEASURES

None of the enteroviruses, including poliovirus, is susceptible to chemotherapy and there are no vaccines, except of course for poliomyelitis. Control therefore depends on hygienic measures, particularly those aimed at reducing the risk of spread by the faecal–oral route. These include not only the proper management of food, water supplies, and sewage, but also individual measures such as handwashing and correct disposal or disinfection of potentially contaminated materials.

5 REMINDERS

◆ The family *Picornaviridae* comprises four genera of **small RNA viruses**, of which two, *Rhinovirus* (Chapter 8) and *Enterovirus*, frequently infect humans.

◆ The picornaviruses are **icosahedral**, about 25 nm in diameter, non-enveloped and have a **positive-sense RNA** genome.

◆ The *Enterovirus* genus comprises the three polioviruses, coxsackieviruses A and B, echoviruses, and enteroviruses. Those infecting humans replicate only in primate cells. They cause a wide range of illnesses, including poliomyelitis, other infections of the CNS, and infections of the heart, skeletal muscles, liver (Chapter 20), skin, and mucous membranes, including the conjunctiva.

◆ Enteroviruses are transmitted predominantly by the **faecal–oral route**, but sometimes via the respiratory tract; conjunctival infections are spread by contact with infective secretions.

◆ In infections of the respiratory tract, gut, and conjunctiva, the viruses invade the mucous membranes directly. In infections of other organs, e.g. CNS and heart, there is a viraemic phase followed by dispersion of virus to the target organs.

◆ Enteroviruses are ubiquitous and **many infections are inapparent**; they affect children more often than adults and are most prevalent in areas where hygiene is poor, or where there is overcrowding.

◆ Laboratory diagnosis involves **isolation of virus** from faeces, throat swabs, CSF, or vesicle fluid. Identification requires neutralization tests or, in the case of coxsackieviruses, inoculation of newborn mice. **IgM antibody capture tests** of the ELISA type may also be useful.

◆ Prevention of poliomyelitis is accomplished by immunization with **inactivated polio vaccine (IPV)** or **live attenuated oral polio vaccine (OPV)**. Both are highly effective, but OPV has a number of advantages, including ease of administration and low cost.

◆ There is no specific antiviral therapy for enterovirus infections and no vaccine other than polio vaccine. Prevention thus depends on **hygienic measures**, particularly those designed to block faecal–oral spread.

HERPESVIRUSES

1 INTRODUCTION

Herpetologists are not virologists who specialize in herpes infections, they are in fact experts on reptiles (Greek, *herpeton*). The name herpes seems to have become attached long ago to this group of viruses because of a rather fanciful idea of the creeping nature of the lesions caused by some of them. The name is, however, appropriate in a more modern context, because these viruses are excellent examples of 'creepers' (Chapter 4).

The herpesviruses form a large and most important group of infective agents, and affect many species both of warm- and cold-blooded animals. In

essential for the activation of certain antiviral drugs (Chapter 29).

2.5 *Antigens*

As might be expected from the many polypeptides contained in these viruses, antigenic analysis and the study of serological cross-reactions are difficult. Conventional tests such as neutralization and complement fixation identify the subfamilies, but for distinguishing between HSV-1 and HSV-2 cross-absorptions are necessary; for such identifications, monoclonal antibodies and restriction endonuclease analysis of viral DNA are more efficient. In the diagnostic laboratory, the antigens of EBV have received most attention since their pattern can give useful information about the stage of infection.

humans they cause a wide range of syndromes, varying from trivial mucocutaneous lesions to life-threatening infections; they may also be implicated in certain cancers. An important property of all herpesviruses is their ability to cause **latent infections** that may subsequently become **reactivated**. They occupy a great deal of medical time both in the clinic and in the laboratory and thus merit detailed attention.

Because this chapter is longer than usual, there are 'Reminders' at the end of each main section.

2   PROPERTIES OF THE VIRUSES

People with eczema are particularly liable to acquire skin infection with HSV, an unpleasant and occasionally fatal condition known as **eczema herpeticum**.

Genital infections have in recent years attracted much attention in Western countries, where publicity about them in the media has contributed to the realization that gonorrhoea and syphilis are not the only sexually transmitted diseases. Nevertheless, their existence has been known for many years.

Because most people have been infected with HSV-1 early in life, genital infection is more often acquired as an initial episode than as a true primary infection. In the male it presents as a crop of vesicles on the **penis**; lesions may also occur within the meatus causing dysuria. **Herpetic proctitis** is sometimes seen in homosexuals. In the female, the lesions are on the **labia**, **vulva**, and **perineum**, sometimes extending to the inner surface of the thighs (Fig. 15.4). **Cervicitis** with vesicular lesions also occurs. **Inguinal lymphadenopathy** is pronounced in both sexes and there is often some fever and malaise. Not infrequently, particularly in male homosexuals, there is also an attack of **meningitis**.

There has been much discussion about the relationship of HSV infection of the cervix and cervical carcinoma. There is certainly a correlation between the incidence of both conditions, but it is by no means certain that HSV is a contributory cause, although some workers claim to have identified HSV DNA sequences in carcinoma cells. Many other factors, such as promiscuity, smoking, and infections with other micro-organisms, in particular papillomaviruses (Chapter 21), may, however, be implicated.

The primary attack is followed by recurrences, which, although less severe, may be frequent. They are the cause of much sexual disability and mental distress.

In children, herpetic vulvovaginitis may indicate sexual abuse, but, needless to say, other causes of infection, such as auto-inoculation from an oral lesion, must also be considered.

Ophthalmic infections are nearly always caused by **HSV-1**. Herpetic **keratoconjunctivitis** is a comparatively frequent infection often characterized by **dendritic ulcers**, so-called because of their branching appearance. The corneal stroma may also be affected (disciform keratitis), and iridocyclitis is a serious complication. Recurrent attacks of keratitis or keratoconjunctivitis, if untreated, may also cause considerable damage to the eye, with corneal scarring and loss of vision.

Meningitis and encephalitis. We mentioned earlier that meningitis occasionally complicates genital tract infections. By itself it is a benign condition; but HSV encephalitis is a serious, life-threatening illness with a high rate of neurological sequelae after recovery from the acute infection. It may occur at any age and is usually caused by HSV-1; there may be concurrent signs of herpes elsewhere, but more often than not it strikes without warning. The onset is often insidious with malaise and fever lasting a few days, followed by headache and

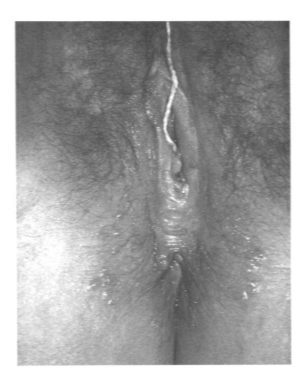

Fig. 15.4 Genital herpes. (Photograph courtesy of Dr David Oriel.)

changes in behaviour. Clouding of consciousness proceeding to coma is a bad prognostic sign. Untreated, the mortality is about 70 per cent with a high rate of neurological sequelae, including mental defect, in survivors (Chapter 24).

Neonatal infections are almost always caused by HSV-2 and are acquired during passage through an infected birth canal. They are described in Chapter 25.

Infection in the immunodeficient host is discussed in Chapter 26. HSV infections, sometimes severe and persistent, are common in patients whose cellular immunity has been impaired by malignant disease or cytotoxic therapy. They are also frequent in patients with AIDS, who may suffer from severe perianal infections.

3.2 *Pathogenesis*

Skin and mucous membranes are the portals of entry in which the virus also multiplies, causing lysis of cells and formation of **vesicles**. In mucous membranes these soon rupture to form shallow ulcers, but in skin they remain intact for several days before crusting over and healing. The clear fluid within the vesicles contains large numbers of virions. In scrapings or sections of vesicles there are cells containing eosinophilic **intranuclear inclusion bodies** (Chapter 4) characteristically surrounded by a clear halo (Fig. 25.7); they represent sites of replication ('virus factories').

Local replication is followed by spread to the regional lymph nodes and then by viraemia, which is of little significance unless there is haematogenous spread to the central nervous system, causing meningitis or even encephalitis (Chapter 24). Such events are, however, comparatively rare; the nervous system is involved by a more subtle mechanism which is central to the pathogenesis of HSV infection.

Soon after replication is under way in the skin or a mucous membrane, virions travel to the **root ganglia** via the sensory nerves supplying the area (Fig. 15.5). Primary infections of the orofacial and genital areas involve, respectively, the **trigeminal** and **lumbosacral** dorsal root ganglia. The virus then becomes **latent** in the ganglia in a way which is not fully understood.

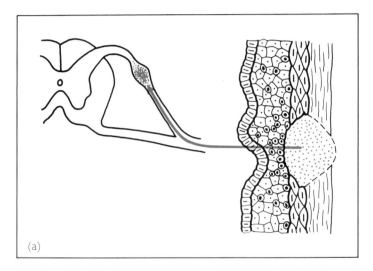

(a)

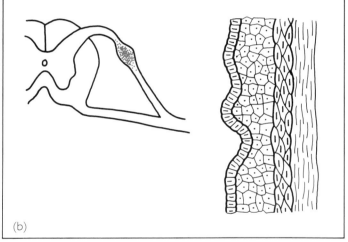

(b)

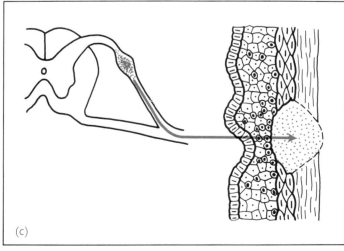

(c)

Fig. 15.5 Latency and recurrence in herpes simplex infections. From the primary lesion, the virus **(a)** travels up the sensory nerves to the dorsal root ganglion and **(b)** becomes latent; **(c)** when reactivated, it returns to the skin by the same route and gives rise to a recurrent lesion.

The simple explanation is that viral DNA becomes incorporated into that of the neurons, but it is difficult to see how this can happen in a non-replicating cell; the alternatives are that some other cell is involved, or that complete virions are able to persist for long periods without replicating. Whatever the answer, once HSV has established itself in a ganglion, **reactivations** are liable to take place at intervals of weeks or months thereafter. Such attacks may be triggered by a variety of stimuli. The popular terms, 'cold sores' or, in the USA, 'fever blisters', illustrate their association with colds and other febrile illnesses; sunlight, menstruation, and therapeutic irradiation are other well-known factors, as is surgical interference with the trigeminal nerve. Again, the mechanism is not clear, but it seems likely that the host cell is stimulated to produce infective virions that pass down the sensory axons to replicate once again in dermal cells, with the production of vesicles at or near the site of the original infection. It may be that those nerve cells in which virus is reactivated become lysed in the process; if so, it is plausible that the eventual destruction of the whole pool of latently infected cells accounts for the cessation of recurrences.

3.3 *Immune response*

IgM and IgG antibodies are produced and seroconversion following a true primary infection is diagnostically useful. IgG antibody persists for many years but does not prevent reactivations, the virus presumably being protected from neutralization within the 'immunologically privileged' site of the nervous system. The antibody response to recurrences is variable, often undetectable by routine methods, and thus of no diagnostic value.

Because of the cross-reactions between HSV-1 and HSV-2, routine serological tests such as complement fixation and neutralization do not distinguish between them. Nevertheless, although infection with one of these viruses usually protects against reinfection with the same variety, it often fails to confer immunity against the other; thus people who have had, say, an HSV-1 infection of the mouth are not fully protected against subsequent genital infection with HSV-2.

The importance of the cell-mediated response in controlling the course of infection is amply illustrated by the tendency of HSV infections to become generalized in some patients with defects in cell-mediated immunity (Chapter 26).

3.4 *Epidemiology*

Most people are infected by the time they are adults. The virus is usually acquired in infancy or childhood but there is, at least in some communities, a second peak at adolescence when opportunities for spread are renewed by sexual activity. Primary infections after the age of 30 years are rare. There are recurrences in about half of those infected but shedding of virus in oral and genital secretions takes place both in them and in those without clinical signs.

3.5 *Laboratory diagnosis*

The clinical appearances of mucocutaneous herpes are usually distinctive enough to make laboratory tests unnecessary. The tests are, however, useful in genital and eye infections, in distinguishing between generalized herpes and varicella-zoster in an immunodeficient patient, and in herpetic meningoencephalitis; the latter topic is dealt with in Chapter 24.

Serological tests, are, by and large, of limited value, especially in diagnosing recurrences. When they are used, complement fixation remains the standard technique in many laboratories but is now being replaced by more sensitive assays such as ELISA. Weak cross-reactions with varicella-zoster are a possible source of error.

Detection of virus rather than of antibody is thus usually the method of choice. In most instances, the distinction between HSV-1 and HSV-2 is academic and there is no need to do the more sophisticated tests needed for identification at this level.

Virus isolation. Scrapings or swabs of active lesions are inoculated into a suitable cell culture—human diploid lines are highly susceptible—and observed for cytopathic effect. Foci of enlarged cells, some of which are multinucleate, appear within 72 h. HSV-2 usually causes more 'ballooning' of the cells than HSV-1 and an experienced observer can often make an accurate guess as to the type of virus. The identity of any doubtful appearances can be established by serial passage, or, much more quickly, by **immunofluorescence staining** or **electron microscopy**.

It must be remembered that, by itself, isolation of HSV from body fluids such as saliva or genital secretions does not prove that it is causing disease, since many apparently normal people shed virus intermittently.

Giant cells and inclusions. Stained smears made from vigorous scrapings of the base of an active lesion often reveal multinucleate cells containing intranuclear inclusions. These appearances may also be observed in sections of biopsy or necropsy material. Such techniques are, however, rarely used nowadays, although they may still be useful in laboratories with limited facilities.

Electron microscopy. If vesicles with clear fluid are present this method is fast and reliable. It is not recommended if there is secondary infection or if crusting has taken place. A drop of fluid is taken by opening the top of a vesicle with a large-gauge needle or the point of a scalpel blade, spreading it on an ordinary microscope slide, and allowing it to dry. In the laboratory the material is taken up in distilled water for negative staining and electron microscopy. The fluid contains numerous virions that are easy to recognize, so that a diagnosis can often be made with a drop only 2 or 3 mm in diameter within an hour of collecting the specimen.

Immunofluorescence. Monoclonal antibodies against both HSV-1 and HSV-2 are now available; these are conjugated with a fluorescent label and used to stain vesicle fluid; by this method, the virions appear in the ultraviolet microscope as bright yellow-green particles. The method has the advantage of immediately distinguishing between the two herpes simplex viruses but, like electron microscopy, demands properly collected specimens taken from suitable lesions.

3.6 *Immunization*

Claims that an experimental vaccine prevents recurrences have not as yet been verified.

3.7 *Treatment*

Acyclovir (ACV) is the drug of choice for treating and, in some situations, preventing HSV infections. Intravenous, oral, and topical preparations are available. The mode of action is described in Chapter 29. In practice, the following general points are important.

As with all antiviral therapy, ACV must be given early to be fully effective. Except in the eye, topical applications are relatively ineffective and the drug is expensive. ACV may, however, be indicated for local infections in immunodeficient patients that would not otherwise need treatment.

ACV is excreted in the urine and the dosage must be reduced in patients with impaired renal function.

Treatment of specific syndromes

Except for severe genital tract infections, primary and recurrent mucocutaneous herpes simplex does not normally need specific treatment. Facial lesions should be kept dry with boric acid or methylated spirit. Long-term continuous oral prophylaxis is very effective in preventing severe recurrent genital herpes but is also very expensive.

Ophthalmic infections are treatable with topical applications either of **idoxuridine** or **ACV**. The efficacy is about the same, but idoxuridine is more liable to induce hypersensitivity. Steroids are strongly contra-indicated.

Encephalitis. A double-blind trial in Scandinavia clearly showed that ACV is superior to vidarabine, both for preventing sequelae and diminishing mortality. Early treatment is particularly important in this condition.

Complications. Bacterial infection of the vesicular fluid leading to pustule formation may occur. In patients with thrombocytopenia the rash may become haemorrhagic and there may be bleeding from body orifices. Postinfectious **encephalitis** may occur, particularly in immunodeficient patients (Chapter 26).

The complications most to be feared are **varicella pneumonitis** and severe multisystem disease, which again mainly affect immunocompromised patients. **Leukaemic children** are at particular risk of pneumonitis, which is often fatal. The term 'generalized varicella' is tautologous, but is used to imply clinically apparent involvement of the viscera, joints, and central nervous system.

Congenital and neonatal infections are described in Chapter 25.

Herpes zoster

'Zoster' is derived from the Latin word for a belt or girdle and refers to the characteristic distribution of the rash when a thoracic dermatome is involved. The attack is heralded by hyperaesthesia and sometimes by pain in the affected area, followed within a day or so by a crop of typical herpetic vesicles which eventually crust over and heal in the usual way (See Fig. 4.1(b)).

Complications. Post-herpetic pain in the affected area is frequent, particularly in the elderly. It usually ceases within a month, but may persist for years. If severe, its resistance to treatment can make life a misery for a minority of patients.

Ophthalmic zoster is a potentially serious complication, when as often happens, the ophthalmic branch of the trigeminal nerve is affected. Both superficial and deep structures of the eye may be involved.

Generalized zoster is similar to generalized varicella, but is distinguished clinically by the presence of vesicles distributed along a dermatome in addition to the more general rash. Encephalitis is a rare complication of herpes zoster.

4.2 Pathogenesis

Varicella is acquired by the **respiratory route**. Because there is no animal model, the pathogenesis has not been verified experimentally, but probably follows the pattern of other acute childhood fevers (Chapter 4). Dissemination of virus in the bloodstream is followed by a rash, at first macular but rapidly developing into papules followed by **vesicles** similar to those caused by HSV and, like them, containing many virions. The lesions become crusted and separate as scabs within 10 days of onset; contrary to early ideas, the scabs are virtually without infectivity.

In some patients the virus becomes **latent** in a dorsal root ganglion. This event is analogous to what happens in HSV infections, but, as one might expect in such a generalized infection, VZV may affect any sensory nerve, those most commonly involved being the thoracic, trigeminal, cervical, and lumbosacral; it is rare for more than one of these to be involved. The proportion of people in which this happens is unknown, because, unlike HSV, latent virus cannot be isolated from the ganglia of cadavers. It can, however, be reactivated, causing an attack of herpes zoster with a characteristic distribution of vesicular lesions along the affected dermatome. Almost invariably, there is only one such episode in a lifetime. Even less is known about the trigger mechanism than in the case of HSV infections, but it is noteworthy that attacks of zoster are especially liable to occur in the **elderly**, in whom immunity is waning, and in **immunocompromised patients**.

4.3 *Immune response*

The antibody response to chickenpox is more clear-cut than in HSV infections. IgM antibody appears within the first week and persists for about 3 months; IgG antibody is detectable for many years. A subsequent attack of zoster results in renewed production of IgM antibody. As in HSV infections, intact cell-mediated immunity is most important in controlling recovery from infection.

4.4 *Epidemiology*

In temperate climates, chickenpox is primarily a disease of young children, but at least two reports show that the age of onset in the Indian subcontinent is much higher, adults being mainly affected with correspondingly more severe infections. The reason for this is not clear.

Since herpes zoster is a reactivation of a latent infection it cannot — or should not — be 'caught' in the same way as varicella. The qualification is necessary because there are reports suggesting that zoster can be acquired by contact with other cases, but this seems inherently unlikely and the evidence is circumstantial. By contrast, **varicella can be readily acquired from patients with an active zoster infection**. This can present a problem in wards containing immunodeficient patients who are particularly liable to develop zoster if they have previously had chickenpox, or who are unduly susceptible to VZV infection if they have not. Furthermore, spread from patients to non-immune nursing staff occurs readily. Management of such situations includes adequate isolation of patients with VZV infections, who ideally should be attended only by nursing staff with antibody to the virus.

4.5 *Laboratory diagnosis*

When laboratory confirmation is needed, it depends on techniques similar to those used for HSV. As we have seen, serological tests are more informative than with HSV, although the results may be blurred by cross-reactions with these viruses. Detection of IgM by radioimmunoassay is useful for diagnosing current or recent infection.

Direct demonstration of the virus by isolation in cell cultures can be a slow business because the cytopathic effect may take up to 3 weeks to become apparent. Under the electron microscope, the virions appear identical to those of HSV; immunofluorescence tests on infected cell cultures will however distinguish between these agents.

4.6   *Active immunization*

A **live attenuated VZV vaccine** is now undergoing trials for preventing chickenpox in immunodeficient children, e.g. those with leukaemia. The initial results are encouraging (see also chapter 26).

4.7   *Passive immunization*

Although there are no really well-controlled trials, **immunoglobulin** prepared from the sera of blood donors with high titres of antibody to VZV seems to be of some use in preventing or modifying severe disease in immunodeficient patients who come into contact with chickenpox or herpes zoster. This preparation (**zoster immune globulin, ZIG**) must be given as soon as possible and preferably within 4 days of contact. It is of no value for treating established infection.

4.8   *Treatment*

Acyclovir is superseding vidarabine for treating severe VZV infections. *In vitro* it is about 20 times less active against this virus than against HSV, but is still effective clinically: for intravenous use, twice the usual dose is given, i.e. 10 mg/kg every 8 h.

Chickenpox in the immunocompetent child does not require antiviral therapy. Calamine lotion containing 1 per cent phenol is useful for soothing the itching. In immunocompromised patients there is always the danger of generalization and varicella pneumonia, for which ACV is useful both in prevention and treatment.

ACV in high dosage can be given during the prodromal stage in an attempt to modify the severity of the infection, but seems to have little or no effect in reducing the incidence of postherpetic pain.

Both ACV and vidarabine are used for controlling herpes zoster in immunocompromised patients; the latter should not, however, be used in the elderly, in whom it may cause Parkinson-like tremor.

4.9   *Reminders*

◆   **Varicella** (chickenpox) is a common childhood infection acquired by the **respiratory route**. The **incubation period** is about 2 weeks.

◆   It causes a generalized **rash**, progressing from macules to papules and then to **vesicles**, similar to those of herpes simplex.

◆   Recovery is usually uneventful, except in **immunocompromised** patients, e.g. leukaemic children, who may develop severe **pneumonitis**.

◆   **Herpes zoster** (shingles) results from reactivation of latent virus in dorsal root ganglia and appears as a crop of vesicles on one or more of the dermatomes of the trunk, or, if the trigeminal ganglion is affected, on the head. **Postherpetic pain** in the affected area may be troublesome, especially in the elderly. Like varicella, zoster may become generalized in immunocompromised patients.

◆   Herpes zoster is infective for non-immune contacts, in whom it may cause chickenpox.

◆   The clinical diagnosis may be confirmed by finding specific **IgM antibody** or by demonstrating virus in vesicle fluid by **electron microscopy** or **immunofluorescence**.

◆   **Passive immunization** of susceptible contacts may be effected with **zoster immune globulin (ZIG)**. An **attenuated live vaccine** shows promise for protecting immunodeficient children.

◆   **Acyclovir** is used for treating severe varicella or herpes zoster in immunodeficient patients.

5   CYTOMEGALOVIRUS (CMV)

The name means 'large cell virus' and derives from the swollen cells containing large intranuclear inclusions that characterize these infections. A similar agent affects guinea-pigs.

5.1   *Clinical aspects*

At first encounter, CMV infections are confusing because they vary so much in terms of age groups affected, mode of acquisition, and clinical presentation. A good aid to memory is to classify them by the ages predominantly affected, since this factor strongly influences the type of disease likely to be encountered (Table 15.2). CMV can be a problem even before the cradle and certainly to the grave.

Table 15.2 Cytomegalovirus infections

Category of patient	Mode of infection
Fetus	From mother, across the placenta
Infant	Contact with maternal body fluids during birth; breast-feeding
Young child	Contact with urine or saliva of other children
Adolescents and adults	Kissing; sexual intercourse; blood transfusion
Transplant recipient	Donated tissues; blood transfusion; reactivation due to immunosuppression

CMV is transmissible to the **fetus** via the placenta, and is an important cause of neonatal morbidity and mortality (Chapter 25).

Normal **infants** can acquire infection from **colostrum** or **breast milk**; maternal antibody does not seem to confer protection. Such infections are usually asymptomatic.

Young children readily become infected, again without overt disease, when they enter crèches or play schools, in which the environment is liable to become contaminated by virus shed in **urine** and **saliva**.

The next wave of infection occurs at **adolescence and early adulthood**, when infection is spread by **kissing** and **sexual intercourse**. At this age, CMV sometimes causes a syndrome like the infectious mononucleosis resulting from infection with EBV, another herpesvirus (Section 6). CMV causes a febrile illness with splenomegaly, impaired liver function — rarely with jaundice — and the appearance of abnormal lymphocytes in the blood; the pharyngitis and lymphadenopathy characterizing EBV infection are, however, absent. A similar illness occasionally follows blood transfusion, the risk being increased when large volumes are used. CMV is found so constantly in promiscuous male homosexuals that, before the discovery of HIV, it was considered as a possible cause of AIDS.

Transplant recipients. The last of our categories is not age-related. The source of infection can be:

◆ **exogenous**, deriving from the donor's tissues or from blood transfusions given in support of surgery; *or*

◆ **endogenous**, i.e. reactivation of an existing infection in the recipient.

Both types of infection, and particularly the second, are facilitated by the associated immunosuppressive treatment. Primary infections are the most dangerous and may result in glomerulonephritis with rejection of a transplanted kidney or in CMV infection of the lungs. This is an **interstitial pneumonitis** with oedema and pronounced cellular infiltration. The mortality is about 20 per cent in bone marrow recipients but only 1–2 per cent in kidney transplant patients. Reactivations are less serious than primary infections and may result in nothing worse than a febrile illness. For further discussion of CMV infections in renal and bone-marrow transplantation see Chapter 26.

5.2 *Pathogenesis*

The pathogenesis of the herpesvirus infections so far described is reasonably straightforward; CMV presents more difficult problems. Infection is acquired by various routes and lasts for life. How this happens is not clear; it may well involve a combination of true **latency**, with integration of viral genome into leukocytes, and **chronic infection** with production of infective virions. The finding of CMV inclusions in renal and salivary cells of people without clinical disease is good evidence for the latter mechanism. Throughout life, there is **intermittent shedding of virus in body fluids**, including saliva, urine, cervical secretion and semen, giving ample opportunities for transmission; transfused blood may also be a source of infection. In most people, the infection is silent, but in overt disease almost any organ is liable to be damaged. As with other herpesviruses, **reactivations** are an important feature of pathogenesis and, to complicate matters further, **reinfections** with other strains of CMV are also possible.

5.3 *Immune response*

Primary infections, but not reactivations, induce **IgM antibody**, which is therefore useful diagnostically. As might be imagined from the persistence of infection, IgG antibodies are present throughout life. Cell-mediated mechanisms are a major component of the immune response.

5.4 *Epidemiology*

The propensity of CMV to be reactivated intermittently and shed in urine and other body fluids, even in the absence of overt illness, makes it a very successful parasite. As might be expected, the opportunities for acquiring CMV are greatest under conditions of poverty, overcrowding, and poor hygiene. In one sense this is an advantage, because infection is acquired early in life when it is usually symptomless. Conversely, the average age of acquisition is greater in communities with high living standards and the outcomes, in terms of the more severe primary infections, are correspondingly worse; this is particularly true of first infections during pregnancy.

5.5 *Laboratory diagnosis*

Serological methods

Tests for **IgM antibody** are best done by radioimmunoassay (RIA), after absorbing out rheumatoid factor which is liable to give false positive results. They can be used on cord blood to test for infection acquired *in utero* and for diagnosing current or recent primary infections in immunocompetent adults. IgM antibody is not induced by recurrent infection or in immunodeficient patients.

Tests for **IgG antibody** include complement fixation, ELISA, and RIA, the latter two being the more sensitive. Such tests are useful for identifying potentially infective blood or tissue donors and for epidemiological surveys.

Detection of virus

CMV can be isolated from **urine**, saliva, and the **buffy coat of heparinized blood**. This virus can be propagated only in human fibroblasts, in which it gives rise to foci of swollen multinucleate cells with characteristic intranuclear inclusions. Such changes may take up to 4 weeks to appear, but infection can be diagnosed as early as 24 h after inoculation by **immunofluorescent staining with monoclonal antibody**.

In histological sections, CMV can be identified by the characteristic enlarged cells containing **'owl's eye' inclusions** (Fig. 15.7). Like other herpesvirus inclusions these are intranuclear, but are more basophilic than those of HSV and VZV. These cytomegalic cells are present in the epithelial or endothelial cells of most viscera. They can also be found in the salivary glands of about 10 per cent of infants dying of unrelated causes. In the kidney, they are seen in the convoluted tubules but not in the glomeruli: the affected cells are shed into the urine, in which they can sometimes be found after centrifugation.

Immunofluorescence with monoclonal antibody can be used to detect virus in tissue sections, but its main value is in staining bronchoalveolar washings for the diagnosis of CMV pneumonitis.

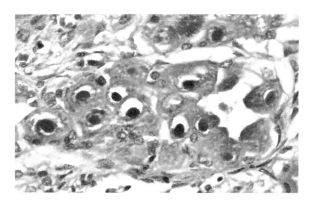

Fig. 15.7 'Owl's eye' intranuclear inclusion of cytomegalovirus.

5.6   *Active immunization*

A live attenuated vaccine is being used experimentally in an attempt to prevent CMV infection in seronegative kidney-transplant recipients; it appears to have a limited success in reducing the severity of infection. In the longer term, such a vaccine might be available for immunizing seronegative women to protect their fetuses, as is done for rubella, but doubts about the possible oncogenicity of such a vaccine will have to be laid to rest before it is brought into large-scale use.

5.7   *Passive immunization*

There is preliminary evidence that the administration of high-titre anti-CMV immunoglobulin protects immunosuppressed patients against CMV pneumonitis.

5.8   *Treatment*

Because CMV lacks the thymidine kinase possessed by HSV and VZV, acyclovir is inactive; an analogue, ganciclovir, has some clinical effect.

5.9   *Reminders*

◆ CMV may be acquired at any age and gives rise to **persistent** infections, during which virus is shed intermittently in **urine** and other body fluids.

◆ In infancy or childhood, transmission is facilitated by poor hygienic conditions. These infections are **silent,** but those acquired later, usually through **sexual activity**, may cause an illness like infectious mononucleosis.

◆ **Transplant recipients** are at particular risk, either from virus transmitted from **donor tissues** or **blood transfusion**, or from **reactivation** of a persistent infection. In these cases, **immunosuppression** is an important factor in increasing the severity of the illness. Primary infection may result in a dangerous **interstitial pneumonitis**.

◆ Diagnosis of **primary** infections depends on demonstrating specific **IgM antibody**, or, by immunofluorescence, viral antigen in human fibroblast cultures inoculated with **urine**, saliva, or **buffy coat** of heparinized blood. In histological sections of kidney, **intranuclear 'owl's eye' inclusions** can be seen in the convoluted tubules.

◆ Since CMV does not code for thymidine kinase, acyclovir has little or no therapeutic effect, although two newer drugs, ganciclovir and 'Foscavir' have some clinical effect.

6   EPSTEIN-BARR VIRUS (EBV)

This human herpesvirus is in some respects the most sinister, for its association with malignant disease is now well established. In 1958, Burkitt described a tumour in African children that occurs in areas with a high prevalence of malaria. He thought that it might be caused by an infectious agent spread by mosquitoes. The mosquito theory was wrong , but 6 years later Epstein and his colleagues discovered a herpesvirus in cultures of the tumour cells. In 1966, American workers showed the association of this virus both with infectious mononucleosis (glandular fever) and with another form of cancer, nasopharyngeal carcinoma, occurring mostly in southern China. In addition to its association with these three syndromes, EBV causes B cell lymphomas in immunodeficient patients. Some species of monkey can be infected experimentally.

6.1   *Virological aspects and pathogenesis*

The reactions of EBV with its host cells differ considerably from those of the other herpesviruses and, in order to make the various syndromes (Table 15.3) more intelligible, will be described before the clinical features.

Normal B lymphocytes placed in cell culture medium die rapidly without dividing. EBV added to the culture **transforms** a proportion of them to blast cells that contain viral genome and continue to divide indefinitely; this process is known as 'immortalization' and has obvious implications for the oncogenic potential of the virus (Chapter 6). A few infected cells go on to produce new virus and are lysed in the process. Like normal lymphoblasts, some EBV-transformed cells secrete **immunoglobulins**. Similar events take place *in vivo* so that, once infected, a person carries B cells containing EBV genome throughout life. EBV induces no cytopathic effect like those caused by other herpesviruses, although its ability to cause inflammatory disease shows its capacity for directly or indirectly damaging cells *in vivo*.

Table 15.3 Epstein–Barr virus infections

Syndrome	Age group mainly affected	Remarks
Infectious mononucleosis	Adolescents, young adults	World-wide distribution
Burkitt's lymphoma (BL)	Children 4–12 years	Endemic in sub-Saharan Africa and New Guinea
Nasopharyngeal carcinoma	Adults 20–50 years	Endemic in southern China
B cell lymphoma	Children and adults	Occurs in some primary and acquired immunodeficiencies

There are two types of the virus, A and B, with differing biological characteristics. Both have a worldwide distribution, but type B appears to have a particularly high prevalence in equatorial Africa and in immuno-compromised patients, including those with AIDS. However, there is as yet no evidence that type B is more oncogenic than type A. Both types may co-exist in the same person.

Replication and induction of antigens

EBV induces virus-specified antigens in B cells that can be detected by immuno-fluorescence staining with the appropriate antibodies. (Fig. 15.8). The first to appear is **EB viral nuclear antigen (EBNA)** which, as its name implies, is found in the nuclei of infected cells. The next is **LYDMA (lymphocyte-detected membrane antigen)** which cannot be stained but is the target for cytotoxic T cells. Another membrane antigen, **'early' MA**, is next, after which either of two pathways may be followed. No further antigens appear in cells destined to become persistently infected; but in a small minority, the successive appearances of early antigen (EA), viral capsid antigen (VCA), and 'late' membrane antigen lead to production of infective virus and cell lysis. For practical purposes you need only remember that the **antibodies to EA and VCA** are sometimes useful for diagnosis (see section 6.2).

6.2 Infectious mononucleosis ('glandular fever')

Clinical aspects

Like CMV, EBV is shed intermittently by a substantial proportion of the population. Infectious virus is generated in the **pharynx**, probably in lymphoid tissue, and appears in the saliva. Thus, like CMV mononucleosis, glandular fever is a 'kissing disease' and the peak incidence is in **adolescents** and

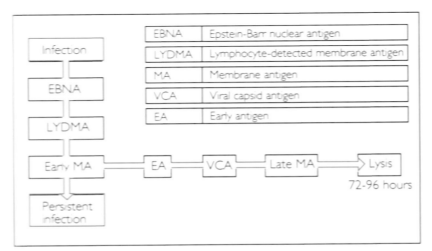

Fig. 15.8 Sequence of appearance of Epstein–Barr virus antigens.

young adults. The incubation period is a month or more. There is **fever**, **pharyngitis**, and **enlargement of the lymph nodes**, first in the neck and later elsewhere. In most patients the **spleen** is palpable and there is some **liver dysfunction**, occasionally with frank jaundice. There may be a transient macular **rash**; it is a peculiarity of the disease that patients given **ampicillin** develop a more severe rash due to the formation by transformed B cells of antibody to this antibiotic.

Complications are relatively uncommon. They include Guillain–Barré syndrome (Chapter 24) or other signs of central nervous system involvement and, rarely, rupture of the spleen. Complete recovery within 3 weeks is the rule but convalescence is sometimes lengthy, with prolonged lassitude and loss of well-being.

Epidemiology

In poor communities EBV, like CMV, is acquired early in life when it causes mainly inapparent infections. Unlike CMV, however, the initial infection confers life-long immunity to subsequent encounters. Infectious mononucleosis is thus most frequent in the developed countries where the infection is acquired later in life, with a peak incidence at 16–18 years.

Laboratory diagnosis

Virus isolation, depending as it does on the transformation of lymphocyte cultures, it impracticable; laboratory diagnosis depends on the haematological findings and on serological tests.

The feature that gives infectious mononucleosis its name is the raised leukocyte count (20×10^9 per litre, 50 per cent of which are lymphocytes). Up to 20 per cent of the lymphocytes are atypical in appearance with bulky cytoplasm and irregular nuclei. These are not, as one might suppose, infected B cells but **T cells** reacting against viral antigens expressed on the B cells and capable of killing the latter.

Heterophil antibodies. B cells transformed by EBV undergo polyclonal expansion, i.e. they produce antibodies with a number of specificities (Greek *hetero*: different). Oddly, one is directed against ampicillin (see 'clinical aspects'). Others agglutinate sheep or horse red cells and are the basis for the **Paul–Bunnell test**. This is done by absorbing the patient's serum with guinea-pig kidney tissue to remove the irrelevant heterophil antibodies of the Forssman type, and then mixing it with the test erythrocytes. In a true positive test, the agglutinating ability can be removed by absorption with ox red cells (Table 15.4). In its original form the Paul–Bunnell test is cumbersome and has been replaced in many laboratories by a **slide agglutination test** ('**Monospot**') based on the same principle. These tests are negative in CMV mononucleosis.

Specific antibodies. Tests of the Paul–Bunnell type may give false negative results in young children. In specialized laboratories, tests for specific EBV antibodies can be undertaken. Test cells expressing the required antigens are stained with

Table 15.4 Paul–Bunnell test for heterophil antibody in infectious mononucleosis

Agglutination of sheep cells after absorbing serum with

Guinea-pig kidney	Ox red cells	Result
Yes	No	Positive
No	Yes	Negative

the patient's serum by the indirect method. Antibody to **EA** indicates a current or recent infection, as does **IgM antibody to VCA**, whereas IgG antibody to VCA is evidence of past infection. These tests are, however, technically demanding and are not in routine use.

6.3   *Burkitt's lymphoma (BL)*

This highly malignant neoplasm (Fig. 4.1 (c)) occurs mainly in **African children** living in the belt where malaria is hyperendemic, roughly speaking between the Tropics of Capricorn and Cancer (Fig. 15.9). The peak incidence is in children 6–7 years old. There is another focus in **Papua New Guinea** and sporadic cases occur elsewhere. It presents as a tumour, usually of the **jaw**, less often of the orbit and other sites. Untreated, it is nearly always fatal within a few months, but is very responsive to **cyclophosphamide**, which, if given early enough, may effect a cure.

EBV genome in circular episomal form can be demonstrated in the tumour cells of most cases of African and New Guinea BL, but not usually in those of the sporadic cases seen elsewhere. All children with African BL possess antibodies to the virus, and there is thus a strong association between the virus infection and the tumour (see also Chapter 6). There is much discussion of the possible role of malaria as a cofactor; its similar geographical distribution is striking, and there is some evidence that its ability to cause immunosuppression interferes with the control by cytotoxic T cells of tumour development, but this is not as yet proven.

6.4   *Nasopharyngeal carcinoma (NPC)*

People in **southern China**, or who originate from there, are subject to an undifferentiated and invasive form of nasopharyngeal cancer, usually presenting as enlarged cervical lymph nodes to which the tumour has metastasized. It is the most common form of cancer in that area, with an incidence of 0.1 per cent of the population. It mainly affects people **20–50 years old**, males preponderating. This neoplasm is also associated with EBV, the evidence being similar to that for Burkitt's lymphoma.

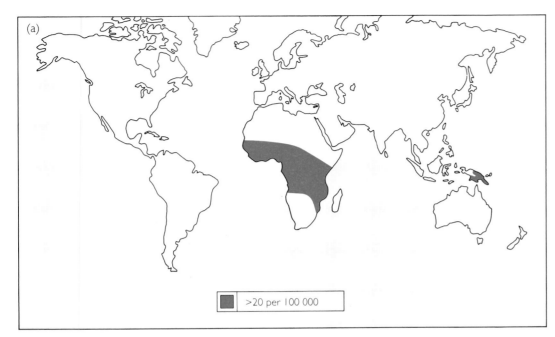

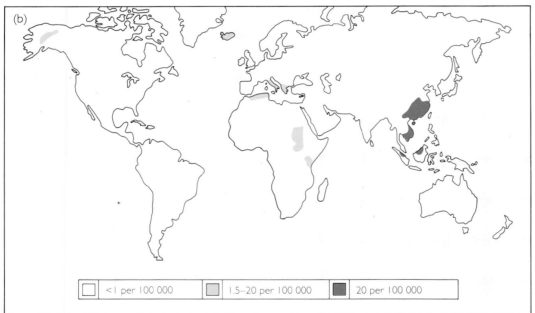

Fig. 15.9 Geographical distributions of **(a)** Burkitt's lymphoma and **(b)** nasopharyngeal carcinoma. (Redrawn, with permission, from Zuckerman, A.J., Banatvala, J.E., and Pattison, J.R. (1987) *Principles and practice of clinical virology*).

In this instance, however, epithelial cells rather than lymphocytes are involved, and the epidemiology is quite different. Not only is the age of onset much later, but there is no association with malaria. Other cofactors appear to operate, one being an **association with certain human leukocyte antigen (HLA) haplotypes**. It has also been suggested that dietary habits may play a part, in particular a high consumption of **nitrosamines** in salted foods. These tumours are relatively inaccessible to surgery or chemotherapy. Even after irradiation, the prognosis is poor.

Probably as a result of failure of T cell control, patients with primary or acquired immune deficiencies are liable to develop **B cell lymphomas** associated with EBV infection (see also Chapters 5 and 26).

For a general discussion of the complex association of virus infections with malignant disease, see Chapter 6.

6.5 *Immunization*

An EBV vaccine might confer enormous benefits in protecting against BL and NPC. An experimental subunit vaccine made from one of the viral envelope glycoproteins has given promising results in tamarin monkeys and may eventually be tested in the field to see if it protects against Burkitt's lymphoma. Nevertheless, the application of such a vaccine on a mass scale is still a long way off.

6.6 *Treatment*

Like CMV, EBV does not code for thymidine kinase and, although acyclovir in high doses has been claimed to have some therapeutic effect in severe infections, the benefit is marginal.

6.7 *Reminders*

◆ EBV, like CMV, may be acquired at any age and causes **persistent infection of B lymphocytes**, in which it induces various virus-specific antigens, of which **early antigen (EA)** and **viral capsid antigen (VCA)** are useful in diagnosis.

◆ **Infectious mononucleosis** (glandular fever) has a peak incidence in adolescents and young adults, in whom it is spread by sexual activity, including kissing. The physical signs include **fever, pharyngitis**, and **enlargement of the lymph nodes and spleen**.

◆ Laboratory diagnosis depends on: (1) the **raised leukocyte count** and **atypical T lymphocytes** in the blood film; and (2) the presence of **heterophil (Forsmann) antibodies** detectable by the **Paul–Bunnell** or **'Monospot'** haemagglutination tests.

◆ As with CMV, acyclovir is of no value in treating acute EBV infections.

◆ **Burkitt's lymphoma** is a malignant tumour occurring mostly in Africa where malaria is endemic, and may be a cofactor. It affects **young**

3.2 *Pathogenesis*

Following an animal bite, rabies virus reaches the central nervous system in man by way of **peripheral nerves**. The virus first replicates in epithelial or striated muscle cells at the site of the bite and gains access to the peripheral

2.2 *Morphology*

The rabies virion consists of a **helical** nucleocapsid contained in a bullet-shaped lipoprotein envelope (Fig. 16.1). Within the lipid envelope are embedded approximately 200 glycoprotein (G) spikes of the virus, responsible for viral attachment to cellular receptors.

nervous system via the **neuromuscular spindles**. The rate of centripetal progress of the virus along the peripheral nerve has been estimated experimentally in mice as 3 mm per day. Once the virus has replicated in the central nervous system it spreads centrifugally along the peripheral nerves to other tissues, including the salivary glands and hair-bearing tissues. In persons infected by aerosols, e.g. in bat-infested caves or laboratory accidents, the virus probably reaches the CNS via nerves supplying the conjunctiva or the upper respiratory tract, including the olfactory nerves.

3.3 *Epidemiology*

Rabies virus is capable of infecting all warm-blooded animals. The reservoirs of infection vary according to the geographical area; dogs and cats are the most important sources of human infection, which is nowadays most frequent in developing countries. The other main reservoirs are wolves in Eastern Europe, the red fox in Western Europe, mongooses and vampire bats in the Caribbean, skunks and raccoons in the USA and Canada, and vampire bats in Latin America. Except for Australia and New Zealand, rabies is present in every continent. The UK has, however, remained rabies-free because of its strict quarantine regulations, which are, of course, relatively easy to enforce in an island. Elaborate precautions have been taken to prevent wild fauna getting to the UK from continental Europe via the Channel Tunnel.

About 700 rabies deaths are reported each year world-wide, but these probably represent only a fraction of the total number. Over 40 per cent of human cases are in children aged 5–14 years; most cases are male, presumably because of greater contact with animals.

The epidemiology of the rabies-related viruses is not clear, but Duvenhage virus, carried by bats, has been recovered from cases of rabies in South Africa, Finland, and the USSR (now CIS).

3.4 *Immune response and laboratory diagnosis*

Little is known concerning T and B cell responses to rabies in humans. A major immune response must be produced to the G protein of the virion since this antigen induces immunological memory in immunized persons.

Fortunately, virus isolation is rarely, if ever, needed for diagnostic purposes. But, if necessary, samples of brain tissue, saliva, CSF, or urine may be injected intracerebrally into newborn mice. The central feature of laboratory diagnosis is demonstration of rabies antigen by immunofluorescence in cells obtained from corneal impressions or hair-bearing skin. This method may also be used on brain smears obtained post mortem from humans or animals. A less sensitive, but useful technique is the search of brain smears for Negri bodies, which are eosinophilic cytoplasmic inclusions. The tissue is taken from the Ammon's horn region of the hippocampus and stained with Mann's stain. Diagnostic procedures for rabies may be undertaken only in specialized, high-security laboratories.

3.5 *Prophylaxis*

The original Pasteur rabies vaccine

Until recently, immunization regimens had changed little from those originally proposed by Pasteur in the late nineteenth century. The strain of rabies virus which Pasteur employed for his now famous immunization experiment was isolated from the brain of a rabid cow. Rabies viruses isolated from the wild are called '**street viruses**'. After passage in the laboratory their virulence is reduced and stabilized. The Pasteur virus was passaged many times intracerebrally in rabbits, by which time the incubation period had become shorter and fixed at 6–7 days; hence the designation '**fixed**' virus. The spinal cords of infected rabbits were dried in air at room temperature, the virulence of the virus contained in them decreasing rapidly with progressive desiccation. In July 1885, a 9-year-old boy, Joseph Meister, was admitted to hospital with multiple and severe bite wounds from a presumed rabid dog. For the next 10 days Pasteur administered a course of 13 injections of rabies-infected cord suspensions, the earliest preparation having been desiccated for 15 days, the subsequent ones for periods decreasing to 1 day. Joseph Meister survived. Pasteur noted, 'The death of this child seemed inevitable and I decided not without lively and cruel doubts, as one can believe, to try in Joseph Meister the method which has been successful in dogs.' A year later Pasteur reported the result of treatment of 350 cases; only one person in this group had developed rabies, a child bitten nearly 4 weeks before treatment commenced. Contemporary figures show that 50 per cent of those bitten should have developed rabies. 'The prophylaxis of rabies is established. It is time to create a centre for vaccination against rabies.' Within a decade there were Pasteur Institutes around the world and, by 1898, 20 000 persons had been treated, with a mortality of only 0.5 per cent. Of course, no one knew about viruses at the time and, since Pasteur was unable to cultivate bacteria from the rabbit spinal cord, he concluded that 'one is tempted to believe that a microbe of infinite smallness having the form neither of bacillus nor a micrococcus is the cause'.

Rabies vaccines derived from infected animal nervous tissues continue to be widely used today (Table 16.2) but it was recognized that neurological reactions in one in every thousand or so recipients limited their acceptability. In the 1950s a rabies vaccine grown in duck embryos also became available and replaced nervous tissue vaccine in the USA and some European countries; it was, however, relatively ineffective and is no longer used.

Table 16.2 Historical development of human rabies vaccines	
Date introduced	**Vaccine and comments**
1884	**Pasteur vaccine**. Historical interest only
1911	**Semple vaccine**. Phenol-inactivated virus vaccine prepared in brains of rabbits, sheep, or goats. Widely used in developing countries. Cheap. Liable to cause neuroparalytic reactions
1957	**Duck embryo vaccine**. Virus inactivated with β-propiolactone. Free from neural tissue, but caused allergic reactions and was a relatively ineffective antigen. Discontinued
1964	**Human diploid cell strain virus (HDCS)**. Virus inactivated with β-propiolactone. Few side-reactions. The most widely used vaccine in developed countries. Expensive

Modern rabies vaccines

A major advance in rabies vaccines was cultivation of the virus in **human diploid cells**. This method provided a potent vaccine which is considerably less reactogenic than those preceding it. HDCS vaccine (Table 16.2) has become the vaccine of choice for prophylactic and therapeutic use. The vaccine is given **prophylactically** to veterinary surgeons, animal handlers, or others at risk from rabies in three doses 1 month apart with a booster at 2 years.

If a person is unfortunate enough to be bitten by a suspected rabid animal, the wound should be thoroughly washed with soap and water, alcohol, iodine, or preferably a quaternary ammonium compound (to which rabies virus is particularly susceptible). Appropriate anti-tetanus treatment should be given to those not immunized against this infection within the past three years. The management then follows the scheme in Table 16.3. A full course of HDCS vaccine consists of 6 doses given on days 0, 3, 7, 14, 30, and 90 days after exposure.

Rabies is one of the few diseases in which vaccine is effective when administered during the incubation period. Over 1.5 million people are still immunized yearly throughout the world, often with the Semple-type vaccines which are given as 21 daily injections subcutaneously. Side-reactions, particularly encephalomyelopathies, are not uncommon with such vaccines.

3.6 Principles of rabies control in animals

Domestic animals

Essential features of a control programme in a rabies area are the **removal of stray animals** and **vaccination** of all domestic dogs and cats. As a result of these procedures, rabies in dogs has decreased dramatically in the USA. Immunization of cats is now being encouraged. All cats and dogs in the USA are immunized with attenuated vaccine commencing at 3 months of age.

Table 16.3 Treatment of people in contact with animals with suspected or actual rabies

Nature of contact	Status of animal*	Treatment of patient†
Indirect contact only	Appears healthy or has signs suggesting rabies	Not needed
Licks to skin	Appears healthy or has signs suggesting rabies	
	(a) under observation for at least 10 days after the contact	Start vaccine immediately. Stop only if animal normal 10 days after the contact
	(b) escaped	Full course of vaccine immediately
	(c) killed	Start vaccine immediately. Stop only if laboratory tests on animal for rabies are negative
Bites	Schedules as for 'Licks to skin' *plus* human rabies immunoglobulin (HRIG), 20 international units/kg body weight, of which half is injected around the bite(s) and half is given intramuscularly	

* Note that the risk of contracting rabies from a wild animal is significantly greater than from domestic dogs and cats.
† Regardless of any pre-exposure prophylaxis, persons exposed to a significant risk of contracting rabies should be given at least two doses of HDCS rabies vaccine.

Any domestic animal that is bitten and scratched by a bat or by a wild carnivorous mammal is regarded in the USA as having been exposed to a rabid animal and unvaccinated dogs and cats are destroyed immediately or quarantined for 6 months. Vaccinated animals are revaccinated and confined for 90 days.

Wildlife

The control of rabies in wild life is a difficult and, some might say, an impossible task. In the USA, continuous trapping or poisoning as a means of rabies control is not recommended, but limited control is maintained in high-contact areas such as picnic grounds. Similarly, bats are eliminated from houses.

In Europe, attempts are being made to control rabies in foxes by a unique immunizing technique. Chicken heads are impregnated with live attenuated rabies vaccine and tetracycline and dropped by helicopter into remote mountainous areas. The foxes eat the heads and become infected and hence vaccinated with the attenuated rabies strain; concomitantly, the tetracycline is deposited in their bones. A simple fluorescence test on the bone tissue of captured foxes can detect tetracycline and indicate whether they have been immunized. By this method, rabies has been eliminated or drastically reduced in parts of Switzerland, and many other European countries are now trying the system.

The natural reservoirs of infection of these viruses are in animals, birds, and even reptiles, among which they are transmitted by the bites of blood-sucking arthropods, usually mosquitoes, sandflies, or ticks. Man is infected only if he gets in the way of their natural cycle by entering areas where they are prevalent and then getting bitten by an infected arthropod. Infection is often inapparent or trivial, but some of these viruses can cause very severe, even fatal illnesses including:

◆ **febrile illnesses, often with rashes and arthritis;**
◆ **infections of the central nervous system;**
◆ **haemorrhagic fevers;**

RNA genome is negative-sense and divided into three segments in the form of a helical nucleocapsid. As might be expected from this segmentation, genetic reassortment can easily be demonstrated in the laboratory, but we do not know whether this is of significance in natural infections. Replication resembles that of influenza viruses but differs in that bunyaviruses mature by budding from the Golgi apparatus and endoplasmic reticulum into intracellular vesicles, after which the cell disrupts to release infective virions.

2.3 Orbiviruses

These agents are a genus of the *Reoviridae* (Chapter 11) and take their name from their large round capsomeres; they contain a segmented **dsRNA** genome.

3 PATHOGENESIS AND IMMUNITY

After the virus is introduced into a subcutaneous capillary by the bite of an infected arthropod there is an incubation period of a few days during which it replicates in the lymphatic system and endothelium; the first signs of illness are usually malaise and fever caused by the subsequent viraemia. Characteristically, signs of infection of the target organs follow 4–10 days later, resulting in a **biphasic** illness. Immunity to reinfection is mediated by the antibody response, which may also confer some protection against related viruses.

4 REMINDERS

◆ The genus **Lyssavirus** contains rabies and three other viruses that infect vertebrates. The viruses are bullet-shaped and the glycoprotein spikes project from the lipid envelope. The genome is **ssRNA of negative polarity**.

◆ The natural reservoir of rabies is the wild animal population and the animal involved varies in different continents. In Europe **domestic dogs** are an important source of infection for humans, whereas the main reservoir in the wild is in **foxes**.

Table 17.1 Arboviruses that affect humans

Family	Genus	Number of serotypes	Percentage pathogenic for man
Togaviridae	Alphavirus	25	30
Flaviviridae	Flavivirus	60	50
Bunyaviridae	Bunyavirus Nairovirus Phlebovirus Hantavirus	200	20
Reoviridae	Orbivirus	100	2

NB. The figures are approximate.

2 PROPERTIES OF THE VIRUSES

2.1 Togaviruses and flaviviruses

Classification

The family *Togaviridae* (Table 17.1) derives its name from the closely fitting envelope surrounding the virions; it contains four genera, of which only the alphaviruses concern us here. The others are rubivirus, which is not arthropod-borne and which causes rubella (Chapter 18), and pestivirus and arterivirus, which infect animals but not man.

The family *Flaviviridae* (formerly classified with the Togaviridae

EPIDEMIOLOGY OF ARBOVIRUS

The development in 1937 of a highly effective **attenuated yellow fever vaccine** by Theiler and Smith was a landmark in the history of immunization. This vaccine, prepared from the 17D strain in chick embryos, has had a major impact on the incidence of urban yellow fever. The immunity induced is very durable, perhaps life-long. The vaccine is not heat-stable and an extensive 'cold chain' (i.e. refrigeration during transport and storage) is needed; this is often a problem in tropical countries and a compromise now often reached is to introduce mass vaccination only when an outbreak is spotted in an urban area.

Both preparation and administration of yellow fever vaccine are strictly controlled and limited to government-approved centres. Apart from use in the face of an epidemic as described above, people who should be immunized include:

◆ **those living or travelling in endemic areas, including tourists**;
◆ **laboratory staff working with the virus**.

Vaccinees must have an International Certificate of vaccination signed and stamped at the approved vaccination centre. It is valid for 10 years. Persons arriving from an endemic area in a country free from yellow fever need such a certificate.

Effective vaccines against several encephalitis viruses are also available in countries where they are prevalent, notably Japan and the United States, but it is clearly impossible to eliminate all the vectors or to provide vaccines against the hundreds of different viruses implicated in these infections. Thus, despite these important but limited successes, it looks as though arboviruses will continue to plague mankind for the foreseeable future.

8 REMINDERS

◆ Members of four families contain arboviruses pathogenic for man: the *Togaviridae*; *Flaviviridae*; *Bunyaviridae*; and *Reoviridae*. Of these, the *Flavi-*

◆ Many infections are **inapparent**, but some are more severe, resulting in:
1. **fever/rashes/arthritis/myositis**; *or*
2. **meningitis/encephalitis/encephalomyelitis**; *or*
3. **haemorrhagic fevers**.

◆ Two dangerous flavivirus infections to be noted particularly are:
1. **yellow fever**, a haemorrhagic fever affecting the liver and other viscera, and its two epidemiological forms, **urban** and **sylvan**;
2. **dengue** and its **haemorrhagic shock syndrome**, which probably results from immunopathological damage due to consecutive infections with different serotypes.

◆ Because these viruses are dangerous they can be handled only in high-security laboratories. **Laboratory diagnosis** is therefore usually based on **serological tests**.

◆ **Control** depends, when practicable, on **vector control** and on **immunization**. Highly effective vaccines are available for some infections, notably **yellow fever** and some of the encephalitides.

RUBELLA: POSTNATAL INFECTIONS

1 INTRODUCTION

Rubella is derived from the Latin *rubellus*, meaning 'reddish', and refers to the pink rash that is seen in most patients. The reason for its popular name, german measles, is obscure; it was written about by German doctors in the last century, but otherwise has no special Teutonic associations. Rubella is predominantly an infection of children, in whom it causes a mild febrile illness. Were this all, we should spend but little time on it, but, as a potent cause of fetal abnormality, rubella has a major claim on our attention. In this chapter we shall deal with the virus itself, with postnatal infections and with immunization; intrauterine infections are discussed in Chapter 25.

2 PROPERTIES OF THE VIRUS

2.1 *Classification and general features*

Rubella virus is enveloped, has a single-stranded, positive-sense RNA genome, and belongs to the family *Togaviridae* (Chapter 17). Because it is the only togavirus that is not arthropod-borne, it has a genus all to itself, *Rubivirus*. This classification is supported by its lack of serological cross-reactions with other togaviruses; but in size, morphology, and mode of replication it resembles them (Fig. 18.1).

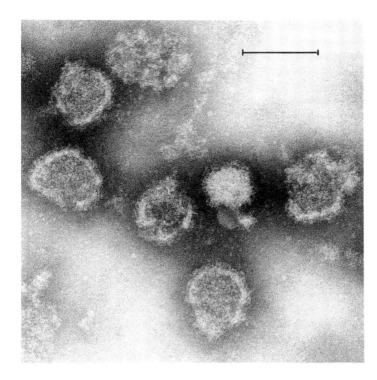

Fig. 18.1 Electron micrograph of rubella virus. (Courtesy of Dr David Hockley.) Scale bar, 100 nm.

2.2 Main antigens

Rubella virus contains complement-fixing antigens and a haemagglutinin. There is very little antigenic variation between strains and, for practical purposes, there is only one serotype, an important consideration both in its epidemiology and in the making of vaccines.

2.3 Pathogenicity for animals

Rubella does not occur naturally in animals other than man, thus differing radically from other togaviruses. Various species of monkey and rodents can be artificially infected, but there is no satisfactory animal model for intrauterine infection.

2.4 Propagation in cell cultures

Rubella virus replicates in a variety of cell cultures, often without inducing cytopathic effects (CPE); in such infected cells, the presence of virus is, however, demonstrable by their ability to resist superinfection with another virus, e.g. an enterovirus. CPE can be induced in a line of rabbit kidney cells (RK13) and in certain others if the conditions are carefully controlled.

3 CLINICAL AND PATHOLOGICAL ASPECTS

3.1 *Clinical features*

Rubella in children

The onset is marked by slight malaise, a small rise in temperature (usually to less than 38°C), and sometimes suffusion of the conjunctivae. Shotty enlargement of the **lymph nodes** in the suboccipital region, behind the ears, and in the neck and axillae is characteristic. The **rash** usually appears a day or two after the onset of symptoms, but is by no means a constant finding. It consists of small pinkish **macules**, rarely exceeding 3 mm in diameter, and usually more discrete and regular in appearance than the eruption of measles. The face and neck are first affected, followed by the trunk. Circumoral pallor is sometimes present, but is not diagnostic. Unlike the pronounced enanthem seen in measles, there may be some very small erythematous spots on the soft palate, but often the buccal mucosa appears normal. A purpuric rash has been described, but is rare.

The vast majority of patients get better within a few days. Post-infection **encephalitis** is rare, affecting one in several thousand patients, and the prognosis is generally good.

Rubella in adults

The main difference between rubella in children and in adults is that in the latter, **polyarthritis** is a not infrequent complication. It most often affects the hands and wrists, but may also involve the larger joints of the limbs. There may also be some myalgia. Rubella arthropathy predominantly affects postpubertal women. It usually clears up fairly quickly, but may persist for months or even years.

3.2 *Pathogenesis*

The portal of entry is the respiratory tract; subsequent spread follows the pattern described in Chapter 4, Section 4.4. The incubation period is usually 14–16 days but may range from 10 to 21 days.

3.3 *Immune response*

Specific IgM antibody appears within a few days of the rash, and is followed soon after by IgG (Fig. 18.2). The titre of IgM increases rapidly, reaching a peak about 10 days after onset and thereafter declining to undetectable amounts over several weeks or months. The rapid appearance of specific IgM

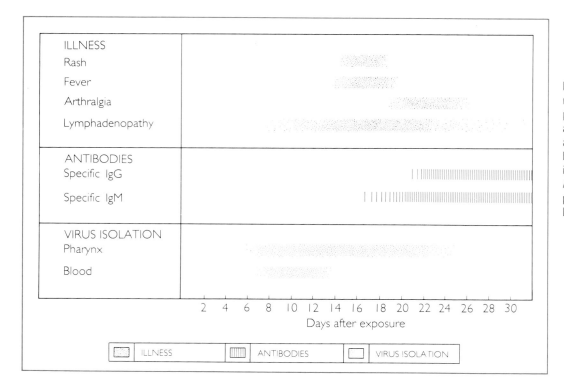

Fig. 18.2 The course of postnatal rubella. (Adapted, with permission, from Banatvala, J.E. and Best, J.M. (1984), in Topley and Wilson's Principles of bacteriology, virology and immunity (7th edn) ed. G. Wilson, A.A. Miles, and M.T. Parker, Vol. 4, pp. 271–302. Edward Arnold, London.)

antibody is invaluable for diagnostic purposes. IgG antibody peaks at about the same time as IgM, but persists for many years.

The cell-mediated response precedes the appearance of antibody by a few days, reaches a peak at about the same time, and persists for many years.

3.4 *Epidemiology*

Because rubella is of significance mainly in relation to fetal infection, an important epidemiological aspect is its prevalence in women of childbearing age. Rubella is nearly always acquired in childhood and appears to be present in all countries, periods of endemicity alternating with epidemics at irregular intervals. Its distribution varies from country to country. Serological surveys (see Chapter 7) show that, in most populations, over 70 per cent of women possess antibody, but in others, where vaccine is little used, notably rural areas in some Third World countries, the figure is lower. This finding might explain the lower than average prevalence of rubella antibody in Asian women now living in London (see Chapter 7, Section 5.2).

Effective immunization programmes radically alter the prevalence; thus in the USA, rubella is now a rare infection.

4 LABORATORY DIAGNOSIS

Because acute rubella is often difficult to diagnose on purely clinical grounds, laboratory diagnosis assumes considerable importance. Tests for rubella infection, past or present, can be considered under three headings:

◆ **screening tests** for rubella antibody, to ascertain the immune status of women of childbearing age;

◆ tests for **acute infection in pregnancy**;

◆ tests on infants for **congenital infection**.

We shall deal here only with the first category. Infections of the pregnant mother and fetus are discussed in Chapter 25.

Screening tests on women. In the UK and elsewhere, women attending antenatal clinics are routinely screened for rubella antibody. If the result is negative, immunization soon after the birth is advised in order to protect subsequent pregnancies. Screening is also advised for women of childbearing age at particular risk of infection, e.g. teachers, and clinic and hospital staff in contact with children. Such tests should be done irrespective of a history of past infection or immunization, which may be unreliable. Since many sera must be processed, the requirement is for a reliable test that can be done on a mass scale and that is not too labour-intensive. Some laboratories use **single radial haemolysis (SRH)** in which many sera can be tested simultaneously in a large gel plate containing both red cells sensitized with rubella antigen and complement. Plates are read after overnight incubation. This method is now giving way to **ELISA** tests for IgG antibody. (See Chapter 27 for details of these tests.)

5 IMMUNIZATION

Rubella vaccine is the only one not given primarily for the benefit of the recipient, but to protect another, in this case a fetus. As we shall see in Chapter 25, the risk of fetal malformation due to rubella acquired during the first trimester is very high.

5.1 *Rubella vaccine*

Rubella immunization was started in the USA during 1969 and in the UK a year later. All the vaccines are prepared from one or another live strain attenuated by repeated passage in cell cultures. Those used in the UK are the Cendehill or RA27/3 strains. Both are very effective immunogens; the latter has the advantage of inducing IgA antibody in addition to IgG, which may be effective in blocking the entry of virus via the respiratory mucosa.

5.2 *Immunization policy*

Until recently, the policy in the UK differed from that in the USA. In the latter, young children of both sexes were immunized, the idea being to protect pregnant women by reducing the circulation of virus in the community, thus minimizing their chances of infection. Clearly, this strategy demands a very good acceptance rate, which was in fact secured by a high level of 'vaccine-consciousness' in the American population, backed up by the requirement for a certificate of immunization as a condition of school entry.

In the UK, the national programme was initially aimed at females only, in the first instance schoolgirls entering puberty, and, later on, seronegative women of childbearing age in contact with children, e.g. health care workers and teachers. This approach was determined by several factors. First, it was held, not unreasonably, that free circulation of virus in the prepubertal child population would provide natural immunity for most females by the age of puberty, and that such immunity might well be firmer and more durable than the protection provided by a vaccine; indeed, it was a possibility that women possessing only vaccine-induced immunity might need a booster later in life. Second, the official policy might well have been conditioned by pessimism about the prospect of securing the necessary high immunization rates, given that in the UK only about 50 per cent of infants were being vaccinated against measles.

There is no doubt that both policies reduced the incidence of congenital rubella, but less so in the UK than in the USA. Figure 18.3 (a) shows the antibody profiles by age group of people in a UK population during the period 1969–85, rubella immunization of teenage girls having been introduced in 1980. The antibody profiles are similar in pre- and postvaccination years. Figure 18.3 (b) shows a similar analysis of a population in Finland, which in 1982 adopted a policy more like that of the USA, vaccine being given to both boys and girls aged 14 months and 6 years. Note the sharp increase in seropositivity in the younger age groups from 1984 onwards. In 1988 a combined measles–mumps–rubella (MMR) vaccine was introduced into the UK and the policy is now, in outline, as follows.

◆ **Boys and girls aged 13–15 months**. MMR vaccine.

◆ **Boys and girls aged 4–5 years**. If no documentary evidence of previous vaccination, MMR before entering primary school (i.e. at same time as diphtheria–tetanus–polio booster).

◆ **Girls aged 10–14 years**. Continue with single rubella vaccine, pending elimination of rubella.

◆ **Non-immune women, before pregnancy or after delivery**. Single rubella vaccine.

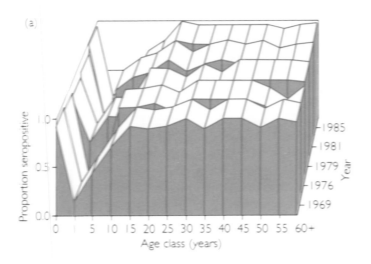

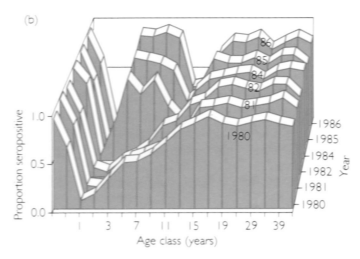

It is now accepted that the immunity resulting from vaccination, although not as solid as that following natural infection, is sufficiently durable to allow a programme of this type. Its aim is the eventual elimination of rubella, measles, and mumps from the UK. Now that the acceptance rate has reached 90 per cent, this goal is well within sight.

Fig. 18.3 The effects of different immunization policies on herd immunity to rubella. For explanation, see text. (Adapted with permission from Anderson, R.M. et al. (1990). *Lancet* **335** 642.)

5.3 *Contra-indications to rubella vaccine*

The main contra-indications are those applying to other live vaccines, i.e. a febrile illness, allergy to one of the constituents (rare), and defective immunity. The important additional contra-indication is **early pregnancy**; women are advised not to become pregnant from a month before to a month after receiving rubella vaccine. That said, no instance of fetal damage has been recorded following inadvertent vaccination shortly before pregnancy or during the first trimester, although serological evidence of infection *in utero* has been obtained in a small proportion of the babies going to term. By and large, there is not a strong case for terminating a pregnancy if rubella vaccine is given inadvertently.

5.4 *Reinfections*

Reinfections, both after second attacks of rubella and immunization, have been recorded and confirmed serologically; as far as is known, such episodes, if in early pregnancy, pose little risk to the fetus, probably because they are not accompanied by the viraemia characteristic of primary infections.

6 REMINDERS

◆ Rubella virus is an **enveloped ssRNA virus** belonging to the *Togaviridae*. It does not naturally infect species other than man. The virus causes cytopathic effects in some cell lines, but isolation is of limited value for diagnostic purposes.

◆ Rubella (german measles) is a world-wide infection of childhood. It is spread by the **respiratory route** and causes a mild febrile illness with **lymphadenopathy** and **rash**. In older people, especially women, it is liable to cause **arthropathy**. It is important because **infection during early pregnancy damages the fetus** (Chapter 25).

◆ For this reason, efforts are made to prevent infection of women of childbearing age by: (1) **serological screening** of certain categories of adult women and immunization of those found to be susceptible; and (2) **mass immunization** of the child population with measles-mumps-rubella (MMR) vaccine.

CHAPTER NINETEEN

SOME EXOTIC AND DANGEROUS INFECTIONS : FILOVIRUSES AND ARENAVIRUSES

1 INTRODUCTION

Most of the viruses described in this chapter have been isolated and characterized in the last two decades and cause haemorrhagic fevers in tropical countries. The common link is their existence in an animal reservoir where the virus may exist quietly until disturbed by human intrusion. They are known as 'exotic' viruses. Some viruses are named after a town or area where an outbreak was first investigated. An example is the filovirus 'Marburg', named after the town in Germany where seven persons died of what was, at that time, a new and unrecorded disease. In this outbreak the common link between the infected persons was the handling of monkeys or monkey tissues. The monkeys had been imported from Africa to provide kidney tissue for preparing poliomyelitis vaccine. Nearly 10 years later an outbreak of an even more lethal haemorrhagic disease caused by a filovirus was described in Zaïre and the Sudan — 'Ebola disease', named in this case after a river in Zaïre. Members of the other viral family described in this chapter, the *Arenaviridae*, also cause outbreaks with disturbing frequency. The filo- and arenaviruses are so dangerous that only high-security laboratories handle them. Hence, 20 years after their discovery, they have still not been as thoroughly studied as some viruses that were isolated much later.

2 THE FILOVIRUSES : MARBURG AND EBOLA

2.1 *Properties of the viruses*

Classification

The family *Filoviridae* is composed of extremely pleomorphic viruses with lipid envelopes, nucleocapsids with **helical symmetry**, and **ssRNA genomes** of **negative polarity**.

Morphology

These viruses have an extraordinary filamentous morphology (Latin *filum*: thread) and sometimes are longer than common bacteria, often with branched, circular and bizarre-shaped forms (Fig. 19.1). Marburg and Ebola viruses can be distinguished from each other by the size of their genomes and their different protein composition; they also differ serologically. Seven virus-coded proteins are detectable in the virions. Beneath the virus lipid envelope a nucleocapsid structure containing RNA can be visualized by electron microscopy.

Genome and polypeptides

The virus genome is **negative-sense ssRNA**, but little is known about the details of viral replication, except that it occurs in the cytoplasm of infected cells. It is assumed that the overall replication strategy is similar to that of other negative-stranded RNA viruses (Chapter 2). The viruses bud from the plasma membranes of infected cells.

2.2 *Clinical and pathological aspects*

Clinical features

The illness caused by the two viruses is very similar, with abrupt onset after an incubation period of 3–16 days. Severe frontal headache, high fever, and back pains characterize the early phase. The patient is rapidly prostrated with diarrhoea and vomiting lasting about a week; conjunctivitis and pharyngitis are usually present. A transient non-itching maculopapular rash may appear after 5–7 days. At this time severe bleeding starts in the lungs, nose, gums, gastrointestinal tract, and conjunctiva in a large proportion of those with clinical illness, preceded and accompanied by thrombocytopenia. Deaths in cases with severe shock and blood loss usually occur between days 7 and 16. The mortality rate may be high, ranging from 25–90 per cent.

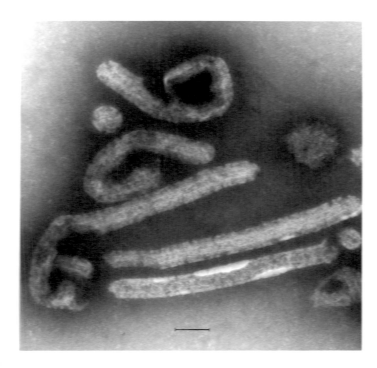

Fig. 19.1 Electron micrograph of a filovirus (Marburg). The virions are highly pleomorphic. (Courtesy of Dr Anne M. Feild.) Scale bar, 100 nm.

Pathology

Both Marburg and Ebola are so-called **pantropic viruses**: they infect and cause lesions in many organs, but especially the liver and spleen which become enlarged and dark in colour. In both these organs severe degeneration and necrosis occur.

Epidemiology

Apart from the first recorded episode, known outbreaks of clinically apparent Marburg disease have been limited to a few individuals in Africa. By contrast, there have been large outbreaks of Ebola infection in the Sudan and Zaïre, where, on the basis of serological surveys, the virus seems to be endemic.

The disease is a zoonosis and although monkeys were involved in the Marburg outbreak, the precise animal reservoir is still unknown. Virus is spread mainly by contact with infected blood although it can be detected in body fluids: in the Marburg outbreak in one case there was laboratory evidence of spread from semen during sexual intercourse. The virus may persist in infected individuals for at least 2 months, which presents an additional transmission hazard. Apart from direct contact with animals, very close contact with an infected patient is a prerequisite for infection, and normal isolation and barrier-nursing methods will prevent transmission.

Laboratory diagnosis

Only certain laboratories have the experience and high-level containment facilities to propagate these viruses. Virus may be isolated from blood and autopsy tissue in Vero cells and identified by its characteristic morphology under the electron microscope or by immunofluorescence. Immunofluorescence of impression smears of infected tissues is also useful. Young guinea-pigs can be infected by intraperitoneal injection and these diagnostic procedures can then be applied to their tissues. A significant rise in antibody is also diagnostic.

Prevention and treatment

No vaccines or specific chemotherapeutic agents are available at present.

3   ARENAVIRUSES

3.1   *Properties of the viruses*

Classification

The family *Arenaviridae* contains a number of viruses (Table 19.1) with a distinctive 'sandy' appearance (Latin *arena*: sand) when virus sections are examined by electron microscopy (Fig. 19.2). The viruses have **ssRNA with negative-sense polarity**.

Morphology

The virions are rather pleomorphic and range in diameter from 50 to 300 nm. Glycoprotein spikes project through the virus envelope and on their inner ends presumably contact the internally situated nucleoprotein. The nucleoprotein has a rather unusual structure of two helical closed circles. The 'sandy' appearance is caused by the inclusion in the virion of cellular ribosomal material during virus budding.

Table 19.1 Arenaviruses that infect humans		
Virus	**Geographical distribution**	**Disease**
Junin	Argentina	Argentinian haemorrhagic fever
Machupo	Bolivia	Bolivian haemorrhagic fever
Lassa	West Africa	Lassa fever
Lymphocytic choriomeningitis	World-wide	Lymphocytic choriomeningitis

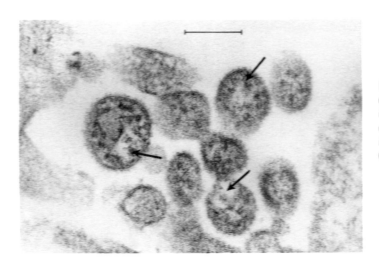

Fig. 19.2 Electron micrograph of an arenavirus (Lassa). This is a thin section of virus in an infected liver. The granules (arrowed) are host-cell ribosomes that have become incorporated in the virions. (Courtesy of Drs David Ellis and Colin Howard.) Scale bar, 100 nm.

Genome and polypeptides

Arenaviruses are unusual negative-stranded RNA viruses in which one of the ssRNA segments is both a positive-strand RNA (at the 5′ end) and a negative-strand RNA (at the 3′ end). Following penetration of the cell and uncoating, the virion transcriptase is activated and copies the predominantly ssRNA into mRNA. Replication is confined to the cytoplasm and budding takes place at the plasma membrane of the infected cells. Presumably it is at this stage that cellular ribosomes are incorporated, by chance, into the budding virion.

3.2   Clinical and pathological aspects

Clinical features

Argentinian, Bolivian and Lassa haemorrhagic fevers. In endemic areas, subclinical infections are frequent. Clinically apparent disease may be severe; the fatality rates in hospitalized patients with Lassa fever range from 15 to 25 per cent. The incubation period is commonly 1–2 weeks and the first signs are non-specific, including fever, headache, and sore throat. The next stage is usually signalled by a rash on the face and neck and a worsening of the patient's general condition. The second week of the illness heralds gastrointestinal and urogenital tract bleeding and a shock syndrome. Even if the patient survives, the convalescence is prolonged with possible hair loss, neurological sequelae, and dizziness. Argentinian and Bolivian haemorrhagic fevers are similar clinically.

Lymphocytic choriomeningitis (LCM). By contrast with the haemorrhagic fevers, LCM is a comparatively mild infection, beginning with headache, fever, and malaise. The illness usually resolves after this stage, but a small proportion of patients develop meningitis or choriomeningitis, which again usually resolves without sequelae (see Chapter 24, Section 2.1); deaths are rare.

Pathology

Extrapolation of data from experimental infections of primates suggests that viral replication in humans occurs in the hilar lymph nodes and lungs following an aerosol infection. The subsequent viraemia causes wide dissemination to other organs including liver, spleen, heart, and meninges. Bronchopneumonia, either primary viral or secondary bacterial, is a common finding. Large amounts of viral antigen are detected in autopsy samples of spleen, bone marrow, and viscera.

Epidemiology

The viruses cause persistent infection in rodents, often without overt signs of disease. Humans catch the disease by contact with their excreta, particularly urine.

Fig. 19.3 **(a)** Prevalence in Africa of human infection with Lassa or related arenaviruses, as shown by serological tests. **(b)** Prevalence in South America of arenaviruses. Only the Machupo and Junin viruses are known to infect humans.

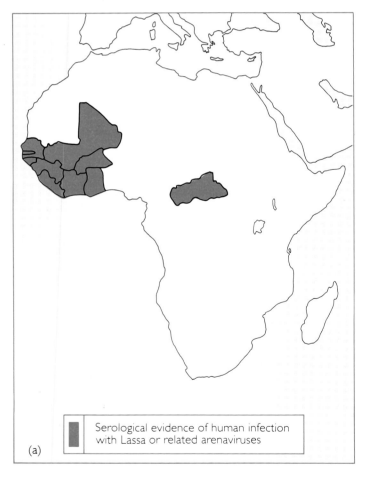

(a)

| | Serological evidence of human infection with Lassa or related arenaviruses |

(b)

| 1 | Tacaribe | Trinidad | 2 | Pichinde | Colombia |
| 3 | Machupo | Bolivia | 4 | Junin | Argentina |

Lassa fever has a certain notoriety because of its first dramatic appearance in the USA in 1969, when, following the death of hospital staff in Nigeria, clinical samples were sent to the world-famous arbovirus laboratory in Yale where they infected and killed a technician and laid low the well known virologist who was head of the unit.

The infection is endemic in rural West Africa (Fig. 19.3), particularly in Nigeria and Sierra Leone, where subclinical infections are frequent, but where there have also been a number of outbreaks with high mortality rates. It is the only viral haemorrhagic fever to have reached the UK, where 10 imported cases have been diagnosed in the last 20 years. Although this number is small, the serious nature of the disease dictates continued vigilance when dealing with febrile illnesses in patients arriving from endemic areas within the previous 3 or 4 weeks.

Argentinian haemorrhagic fever is caused by the Junin arenavirus (Fig. 19.3(b)). Those predominantly infected are male field workers who come into contact with the excreta of chronically infected wild rodents. Thousands of cases may occur when the maize is harvested in late spring and early summer; the case fatality rate is 10–20 per cent.

Bolivian haemorrhagic fever (Fig. 19.3(b)) is due to another arenavirus, Machupo, which again is transmitted from rodent excreta, although person-to-person spread is not unknown. Outbreaks have occurred both in town- and country-dwellers, but the incidence has greatly diminished in recent years.

Lymphocytic choriomeningitis. This virus is widespread throughout the world in domestic house mice, who are chronically infected without clinical signs. Contact with mice or their excreta brings the virus to humans as a 'dead end' infection, with no further spread to others (see Chapter 24, Section 2.1).

Laboratory diagnosis

The first task is to take a careful history so that the patient can be assigned to an appropriate risk category (Fig. 19.4). Since malaria figures prominently in the differential diagnosis, the details of what antimalarial drugs have been taken must be elicited, and blood films, treated with formaldehyde, must be examined for parasites as a routine.

Only specially trained personnel working in high-security laboratories are able to handle clinical specimens with safety (Fig. 19.4). Virus can be recovered from the blood and urine of acutely ill patients for several weeks. Virus may also be isolated from throat swabs and urine and from autopsy specimens of lymphoid tissue, bone marrow, and liver. A single test for specific IgM antibodies provides a useful serological indicator of infection, particularly in Lassa fever.

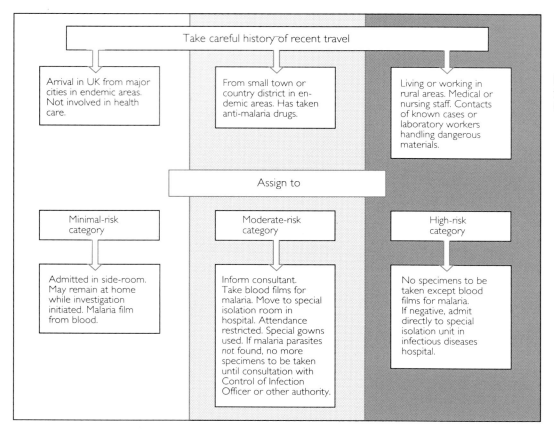

Fig. 19.4 Summary of procedures for dealing with suspected haemorrhagic fevers.

Prevention and treatment

The treatment of choice of Lassa fever is the antiviral drug ribavirin, either alone or with convalescent plasma. It is important that the drug be administered as soon as possible after onset.

For all the haemorrhagic fevers, supportive care is often life-saving, and comprises careful maintenance of fluid and electrolyte balance, protein replacement therapy, and appropriate support for cardiac complications.

Control of rodents is an important method of controlling arenavirus infections, but may be impracticable in rural areas.

4 REMINDERS

◆ The filoviruses **Marburg** and **Ebola** have a **bizarre filamentous morphology**. The **arenaviruses** have a '**sandy**' appearance due to incorporation of cellular ribosomes in the virion. The genomes are all **negative-stranded ssRNA**.

◆ The arenaviruses cause persistant infections in rodents and are transmitted to humans by **contact with rodent excreta**. They are prevalent in Latin America and parts of Africa. Lymphocytic choriomeningitis virus occurs world-wide. The filoviruses may cause zoonotic infections in monkeys.

◆ In humans, infections are often inapparent, but the **filoviruses** and the **Lassa**, **Junin**, and **Machupo arenaviruses** may cause severe **haemorrhagic fevers** with high fatality rates; **LCM virus** may cause **'aseptic' meningitis** or rarely, meningoencephalitis.

◆ Lassa and filovirus infections are **pantropic**, causing lesions in many organs, notably the liver and spleen. Prostration, diarrhoea, rashes, vomiting, and bleeding from the respiratory, gastrointestinal, and urogenital tracts are features of the haemorrhagic forms of these illnesses.

◆ The **diagnosis** is based on the **history** and circumstances of travel in endemic areas, the **exclusion of malaria** and other tropical fevers, and on **isolation of virus** and **serological tests** in a specialist high-security laboratory.

◆ Prompt treatment with **convalescent plasma** and **ribavirin** may be of some value in Lassa fever.

HEPATITIS VIRUSES

1 INTRODUCTION

It is paradoxical that the liver, a site of massive viral replication in the acute infectious fevers, nearly always escapes serious damage during such episodes; this is because the viruses concerned multiply in the Kuppfer cells of the reticuloendothelial system. The targets of the true hepatitis viruses to be discussed in this chapter are the hepatocytes themselves. Nevertheless, liver damage may also be caused by other viruses, which for completeness are listed in Table 20.1.

In the days when syphilis was treated with Ehrlich's 'magic bullet', arsphenamine, patients often became jaundiced, an effect put down to toxicity of the drug for the liver. We now know that the cause was a virus spread from one person to another by inadequately sterilized syringes used to give the injections. Later, it became apparent that there were at least two forms of transmissible hepatitis. One, known as 'serum hepatitis', was spread by injections and particularly by blood transfusions; the other seemed to be acquired by the oral route and was labelled 'infectious' or 'catarrhal' hepatitis. The first type had a long incubation period of 2–5 months, whereas that of the second was shorter, about 2–6 weeks. Neither form was infective for the usual laboratory animals or culture systems and there was confusion about their true nature until experimental inoculations of humans distinguished between the infective agents that we now know as the hepatitis A and B viruses. Since then, other hepatitis viruses have been discovered and are described in Section 4.

2 HEPATITIS A

2.1 *Properties of the virus*

Classification

Hepatitis A virus (HAV) is classified as a member of the *Enteroviridae* and is officially referred to as enterovirus 72.

Morphology

The virions have **cubic symmetry** and are 27 nm in diameter; they resemble those of other members of the family (see Fig. 14.2).

Genome and polypeptides

The nucleic acid is **ssRNA with positive polarity** and a molecular weight of approximately 2.3×10^6. It codes for four polypeptides, VP1, VP2, VP3, and VP4. There is only one serotype.

Replication

HAV cannot be propagated in the laboratory as readily as other enteroviruses. It can be grown in certain monkey kidney and human diploid cells but so far the replication cycle has not been fully worked out; it probably resembles that of the other enteroviruses.

2.2 *Clinical aspects*

The **incubation period** is **2–6 weeks**. Many infections are silent, particularly in young children. Clinical illness usually starts with a few days of malaise, loss of appetite, vague abdominal discomfort, and fever. The urine then becomes dark and the faeces pale. Soon afterwards **jaundice** becomes apparent, first in the sclera and then in the skin; if severe, it may be accompanied by itching. The patient starts to feel better within the next week or so and the jaundice disappears within a month. Complications are almost unknown and the mortality is only of the order of 1/1000.

2.3 *Pathogenesis and pathology*

Although chimpanzees and marmosets can be infected experimentally, the difficulty in isolating virus has not so far allowed detailed study of the pathogenesis. HAV first replicates in the **gut**; there is then a **viraemic phase** during

Table 20.1 Viruses causing hepatitis as part of a general infection

Virus	Remarks
Epstein–Barr virus	
Cytomegalovirus, generalized herpes simplex	Especially in congenital infections
Rubella	Congenital form
Mumps	Rare
ECHO viruses	Rare
Yellow fever	Generalized infection; liver is main target organ

which, presumably, the virus gets into the **liver**. Further replication in the hepatocytes causes necrosis of these cells, which tends to be more pronounced toward the centres of the lobules. There is proliferation of Kupffer and other endothelial cells and secondary periportal inflammation with mononuclear cells.

2.4 *Immune response*

Specific IgM appears during the prodromal phase, is present at high titre by the time jaundice is apparent, and persists for several months. IgG antibody is detectable for many years after infection and protects against further attacks.

2.5 *Epidemiology*

The mode of spread is the same as for other enteroviruses, i.e. transmission is by the **faecal-oral route** (Chapters 4 and 14). Like them, HAV survives for long periods in water and wet environments. Large quantities of virions are excreted in the faeces for several days before and after the onset of jaundice, but after a week the patient's stools may be regarded as non-infectious. Hepatitis A is most widespread in countries where sewage treatment and hygiene generally are inadequate, most persons acquiring it as a subclinical infection in early childhood. It occurs both in endemic form and as epidemics, some of which have been traced to infected shellfish.

2.6 *Laboratory diagnosis*

During the acute phase, liver dysfunction is indicated by **raised serum bilirubin and transaminases** and a depressed prothrombin level. The specific diagnosis is readily made by an **ELISA** or **radioimmunoassay test for specific IgM**. During the early stage of jaundice the virus can be identified in the faeces by immune electron microscopy but, since its presence is transient, this is not a routine diagnostic method.

2.7 *Control measures*

Passive immunization

Individuals travelling from temperate to subtropical and tropical countries can be given an injection of **normal human immunoglobulin** (HIG) shortly before departure. This contains enough anti-HAV antibody to confer sufficient passive immunity to prevent or modify an attack during the next 3–6 months.

Hepatitis A vaccine

Following the preparation in Germany of an effective formalin-inactivated vaccine made from HAV grown in human diploid cells, such a vaccine was licensed in the UK early in 1992 for use in adults.

Hygienic measures

Control of infection in the community depends on maintenance of hygiene, a counsel of perfection that is often very difficult to achieve.

A **food handler** with hepatitis A must be kept away from work for 2 weeks after the onset of jaundice. Fellow workers should **not** be given immunoglobulin as prophylactic since this may merely mask an attack without completely preventing it, an obviously dangerous situation; they should be kept under surveillance and asked to report any illness during the next 12 weeks.

In **hospitals**, patients should be nursed with appropriate precautions against the spread of an enteric infection, with particular attention to safe disposal of faeces during the infective period.

There is no specific treatment for hepatitis A.

3 HEPATITIS B

Although the world-wide incidence of hepatitis A must run into millions, as a cause of serious disease and death this infection pales into insignificance beside hepatitis B. Like HAV, the causal agent of hepatitis B does not infect small laboratory animals, chick embryos, or routine-type cell cultures. The first clue to its nature was obtained in 1964 by Blumberg, who, during a survey quite unconnected with hepatitis, found that an antibody in serum from a haemophiliac precipitated with an antigen in the blood of an Australian Aboriginal. The antigen proved to be a component of the elusive hepatitis B virus (HBV); and the involvement in this discovery of a haemophiliac and an Aboriginal was, as we shall see, highly significant.

3.1 Properties of the virus

Perhaps in no other virus infection is a knowledge of the causal agent and immune responses so important for understanding the clinical and epidemiological aspects and the principles of laboratory diagnosis. Hepatitis B is rather complicated, but a little concentrated study here will serve you well when you meet the infection in practice.

Classification

The family name is *Hepadnaviridae* (*Hepa*titis DNA viruses). As well as HBV, which is found only in humans, it includes at least three similar agents that respectively cause hepatitis in ducks, chipmunks, and squirrels and are thus useful for research.

Morphology

Electron microscopy of the blood of acute and some chronic cases of hepatitis B reveals many particles differing in shape. (Fig. 20.1). One type is 42 nm in diameter, and double-shelled; this is the complete virion, sometimes called the Dane particle after its discoverer. The others are spheres or tubules 20–22 nm in diameter; they consist only of excess surface antigen, i.e. the glycoprotein forming the outer layer of the double-shelled Dane particle. The core of the latter is an **icosahedral** nucleocapsid containing:

◆ the **DNA genome**;
◆ a DNA-dependent **DNA polymerase** involved in replication;
◆ hepatitis B core antigen (**HBcAg**);
◆ hepatitis B e antigen (**HBeAg**).

Genome

The genome consists of circular DNA with a molecular weight of about 2.2×10^6 and a length of 3200 nucleotides. Its conformation is unique in that for most of its length the DNA is double-stranded but one strand ('short') has a gap about 700 nucleotides in length. The 'long' strand has a nick near the 5′ end. (See Chapter 1, Fig. 1.2.)

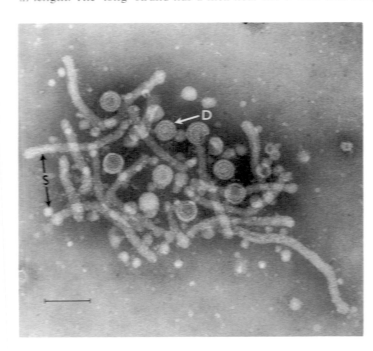

Fig. 20.1 Electron micrograph of hepatitis B virus. D, Dane particles; S, filaments and spheres of hepatitis B surface antigen (HBsAg). (Courtesy of Dr David Hockley.) Scale bar, 100 nm.

Replication is complicated and will not be described in detail. The most significant feature is the transcription of an RNA intermediate from the positive-sense DNA contained in the infecting virion; this is accomplished by a virus-coded reverse transcriptase. The RNA serves as a template for new negative-strand DNA, which in turn is transcribed by a DNA-dependent DNA polymerase to positive-strand progeny DNA for incorporation into the new virions. This **reverse transcription** is similar to what happens with retroviruses (Chapters 2 and 22). The small genome is similar in size to that of HAV but codes for a much more elaborate set of polypeptides by making use of overlapping (or open) reading frames (see Fig. 2.8).

Antigens and antibodies

The core itself has its own antigenic specificity, referred to as HBcAg. The surface antigen is named HBsAg and is in fact the one originally found in the blood of the Aboriginal; it was, and sometimes still is, referred to as 'Australia antigen', but please forget this outmoded term and remember only HBsAg.

HBV markers. The main antigens **HBsAg**, **HBcAg**, and **HBeAg** each induce corresponding antibodies (Table 20.2). With the exception of HBcAg, all these antigens and antibodies, together with the viral DNA polymerase, can be detected in the blood at various times after infection and are referred to as '**markers**', because their presence or absence in an individual patient mark the course of the disease and also give a good idea of the degree of infectivity for others. HBcAg is readily detectable only in the hepatocyte nuclei.

Table 20.2 Serological markers of hepatitis B infection

Marker	Remarks	Present in
Antigens		
HBsAg	Surface antigen, not infective	Acute and chronic infections, including antigenaemia
HBeAg	Found in core of virion. Presence in blood indicates infectivity	Acute and chronic hepatitis
Viral DNA polymerase	As for HBeAg, above	As for HBeAg, above
Antibodies		
Anti-HBs	Indicates recovery; protects against reinfection	Convalescence
Anti-HBe	Presence indicates little or no infectivity	Convalescence
Anti-HBc	In IgM form, indicates recent infection	The first antibody to appear. Persists in IgG form for life

N.B. HBcAg, the core antigen, is not readily detectable in blood and is not used as a marker.

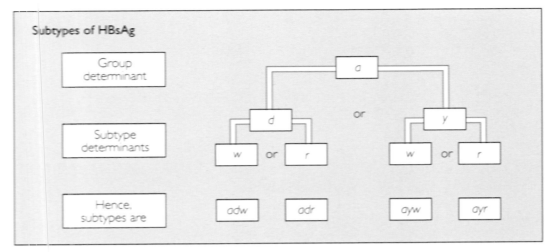

Fig. 20.2 The antigenic subtypes of HBsAg.

HBV subtypes. As well as the main antigens already mentioned, HBcAg, HBeAg, and HBsAg, the surface antigen is endowed with serological specificities that enable us to define subtypes of the virus. There is a group-specific antigenic determinant, *a*, associated with either of two subtype determinants, *d* or *y*. These in turn may be associated with either of two determinants, *w* or *r*. Figure 20.2 shows how the four main subtypes of HBV are derived from these combinations.

These subdeterminants are of no clinical significance, but are useful epidemiologically since their geographical distributions differ. They may also be helpful in deciding whether a particular carrier — e.g. a surgeon or dentist — is the source of infection of another person. The finding of identical subtypes would, of course, not confirm the possibility, but differing subtypes would rule it out.

3.2 *Pathogenesis and pathology*

The mode of transmission is quite different from that of hepatitis A. **HBV is transmitted only in blood and body fluids, including cervical secretions and semen**, but not in faeces or urine (unless of course they are contaminated with blood). It is not transmitted by the respiratory route. HBV is even more difficult to propagate in the laboratory than HAV and many aspects of pathogenesis remain obscure, in particular, what happens to the virus during the long incubation period. Eventually, however, replication starts in the hepatocytes, which seem to be the only host cells. Very many complete virions and even more HBsAg particles are liberated into the bloodstream. The latter contain no DNA and are therefore not infective; but the Dane particles are so numerous — up to 10^{10}/ml — that as little as 0.0001 ml of blood can transmit the infection. This means that minor abrasions or cuts can serve as portals of entry for such minute amounts of infective material. HBV is readily spread by

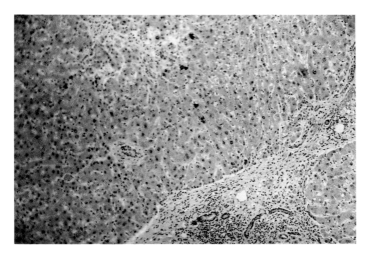 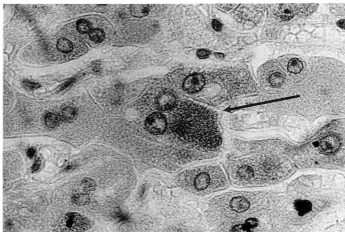

sexual intercourse, particularly among male homosexuals, and in **intravenous drug abusers** by sharing of needles and syringes.

In the acute phase, the pathological changes in the liver (Fig. 20.3(a)) are similar to those seen in hepatitis A (Section 2.3), except that a number of the hepatotocytes are stuffed full of viral surface antigen, which gives them a 'ground-glass' appearance when stained with haematoxylin and eosin. Orcein stains the HBsAg brown (Fig. 20.3(b)). In chronic aggressive hepatitis B (Section 3.3), there may be cirrhotic changes, which, in a proportion of those thus affected, are a precursor of primary liver cancer.

Fig. 20.3 Section of liver from a case of hepatitis B. Orcein stain. **(a)** Low power. Some of the hepatocytes are necrotic and others contain darkly stained HBsAg. Note the dense infiltration with mononuclear cells. **(b)** High power. HBsAg in a hepatocyte (arrowed).

3.3   *Clinical syndromes*

Patterns of infection are much more variable than with hepatitis A and are influenced by age, sex, and the state of the immune system. We shall start with acute infections acquired well after infancy and then consider the special problems posed by perinatal infection.

Postnatal infections

Acute infection may be subclinical, especially in young children or in those with impaired immunity. Typically, however, after an incubation period of 2–3 months, there is a prodromal phase similar to that of hepatitis A, but sometimes marked by a transient rash and arthropathy, probably due to virus/antibody interaction. This is followed by overt jaundice, after which 90 per cent of patients recover uneventfully within a month or so. In others, however, the outcome may be chronic infection or even rapid death (Fig. 20.4).

Figure 20.5(a) shows the course of a typical acute infection in an immunocompetent adult. Even before the appearance of jaundice there is a rise in serum transaminases and HBsAg is detectable in the serum, followed soon afterwards by HBeAg and DNA polymerase (not shown). Anti-HBc is the

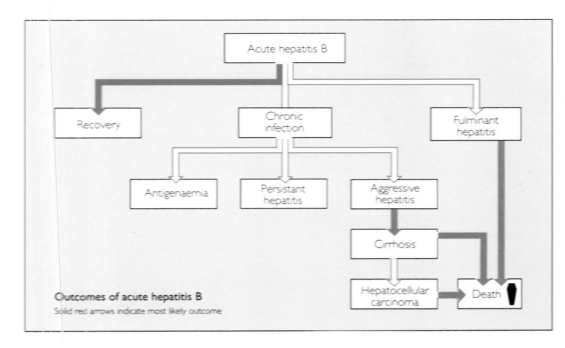

Fig. 20.4 Clinical outcomes of hepatitis B infection.

Outcomes of acute hepatitis B
Solid red arrows indicate most likely outcome

Fig. 20.5 Course of acute and chronic hepatitis B (*below*). **(a)** Acute hepatitis B; **(b)** chronic antigenaemia and persistent hepatitis; **(c)** chronic aggressive hepatitis.

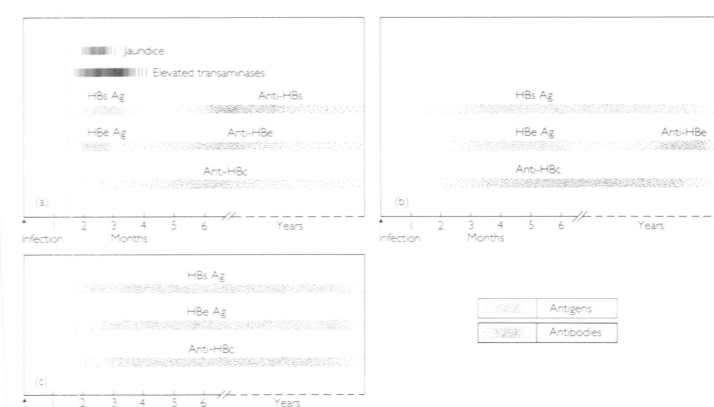

first antibody to appear. The next is anti-HBe, a good prognostic sign since its production heralds the disappearance of HBeAg and thus of infectivity. Although HBsAg is the first antigen to appear, anti-HBs is the last antibody to do so; its arrival indicates complete recovery and immunity to reinfection.

At the other end of the spectrum (Fig. 20.4) one patient in a thousand, usually a female, develops **fulminant hepatitis** and dies within 10 days in hepatic coma; this is the result of an abnormally active destruction of infected hepatocytes by cytotoxic T lymphocytes.

We are now left with the 10 per cent or so of patients who do not come into either of these categories, but who become **chronic carriers**. This diagnosis is made when the serological profile has not reverted to the normal post-recovery pattern within 6 months of onset. There are three main varieties.

In **chronic antigenaemia** the patient fails to form anti-HBs and the appearance of anti-HBe may be delayed (Fig. 20.5(b)). Although HBsAg persists in the blood for many years, liver function is normal, the patient is well and is of little or no danger to others. This picture is often seen in those with impaired immunity. The serological pattern is similar in **chronic persistent hepatitis**, in which, however, there is a mild degree of liver damage.

Chronic aggressive (or active) hepatitis is a different story, and these patients create the great bulk of the problems surrounding hepatitis B. They fail to produce either anti-HBs or anti-HBe (Fig. 20.5(c)). As a result they continue to carry both HBsAg and infectious virions in their blood and **are thus infectious for others**; they are sometimes referred to as 'supercarriers'. There is significant damage to the liver parenchyma and raised transaminase levels (not shown in Fig. 20.5(c)) bespeak impaired function. These patients are liable to repeated episodes of hepatitis and are at risk of developing cirrhosis; some may eventually succumb to malignant disease of the liver.

Hepatocellular carcinoma, or primary liver cancer, arises as a result of integration of viral genome into the DNA of the hepatocytes (Chapter 6). This happens only after a chronic infection with continuing production of complete virions has been in progress for at least 2 years. Cofactors, such as aflatoxin contaminating groundnuts in Africa, have been suggested as contributory causes, but there is now no doubt that HBV itself is the primary carcinogen. The observation that the world distributions of hepatitis B and primary liver cancer are virtually identical (Fig. 20.6) provides strong support for this statement.

Perinatal infections

Infants born to a mother with acute hepatitis B may themselves become acutely infected. By contrast, infants born to mothers who are **HBeAg-positive carriers** do not develop the acute disease. They do, however, stand a 95 per cent chance of becoming infected, either from contact with blood and body fluids at birth or within the next few months by close contact with the mother and siblings. The acquisition of maternal anti-HBc across the placenta seems

Fig. 20.6 Hepatitis B and liver cancer. Geographical distributions of: **(a)** HBsAg carriers (prevalence 5 per cent or more) and **(b)** primary hepatocellular carcinoma (annual incidence 10 cases or more per 100 000 population).

to impair the normal immune response; nearly all such infants proceed themselves to become **HBeAg-positive carriers**, and many die later in life of cirrhosis or liver cancer. The outlook for boys is worse than for girls: 50 per cent of males eventually die of these complications, compared with only 15 per cent of females, who seem to mount a better immune response. The preponderance of surviving females favours perpetuation of the carrier state in populations with a high prevalence of the disease, since it is they who pass the infection to the next generation. Since infection acquired early in life does not present as an acute infection with jaundice, the clinical pictures of hepatitis B in areas of high and low endemicity differ considerably.

3.4   *Immune response*

Although the appearance of anti-HBs indicates recovery, it plays little part if any in that process, which is primarily effected by **cytotoxic T cells** (T_c). We explained in Chapter 5 how T_c lymphocytes act on infected cells only if the viral antigen is presented in association with class I histocompatibility antigen. Hepatocytes are normally not well endowed with this antigen, but HBV infection stimulates production of alpha-interferon, which in turn increases the display of class I antigen on the liver cells and thus permits their lysis by the cytotoxic lymphocytes (Chapter 5).

3.5   *Epidemiology*

The prevalence of hepatitis B in a community can be estimated by the presence of anti-HBs, or, more readily and informatively, by the proportion of the population who are HBsAg-positive carriers. The carrier rates vary widely in different parts of the world.

The **high** prevalence areas (10–20 per cent) are East and South-east Asia, the Pacific Islands, and tropical Africa.

The CIS (ex-USSR), the Indian subcontinent, parts of Africa, Eastern and south-eastern Europe, and parts of Latin America are areas of **medium** prevalence (2–10 per cent).

The prevalence is **low** (< 1 per cent) in the rest of Europe, Australia and New Zealand, and Canada and the USA.

In the UK, the HBsAg carrier rate in the general population is less than 1 per cent; higher rates in the various ethnic minorities reflect to some extent the prevalences in their countries of origin.

Some facts and figures

It is estimated that there are 200 million HBV carriers in the world; of these, 75 per cent were infected at birth.

There are 200 000 new cases every year in the USA, of whom 20 000 become carriers; of these, 4000 eventually die of cirrhosis and 1000 of liver cancer.

The global death rate from hepatocellular carcinoma is estimated at 250 000 per annum.

High-risk groups

This term applies to people who by reason of their country of birth, way of life, or type of work are at higher than average risk of acquiring HBV infection or of passing it on.

We have already described one such group, babies born to carrier mothers in high endemicity areas. Table 20.3 gives examples of high-risk

Table 20.3 Groups at higher than average risk of HBV infection

Category	Risk factors or group
General community	Sexually promiscuous people; intravenous drug abusers; sexual partners of HBeAg-positive carriers; infants of HBeAg-positive mothers
Patients	Repeated blood transfusions; long-term treatment with blood products, e.g. haemophiliacs
Health care staff	Work in mental institutions; tours of duty in high endemicity areas; surgical and dental operations; some pathological laboratory work, including autopsies; work in sexually transmitted disease clinics

groups in areas of low endemicity; all, in one way or another, are liable to come into contact with infective blood or body fluids. At this point, we recall that the first clue to the nature of HBV was the reaction between surface antigen in the serum of an Australian Aboriginal and what we now know to have been anti-Hbs in the blood of a haemophiliac, both of whom would be in high-risk groups.

3.6   *Laboratory diagnosis*

Specific tests

The first test to be performed is for surface antigen. This is normally done by ELISA but, if the result is needed urgently, more rapid tests are available. They might be needed when, for example, a patient from a high-risk group needs emergency surgical or dental treatment and his or her carrier status must be determined to protect the operator. They can be done by **reverse passive haemagglutination**, in which commercially available erythrocytes coated with anti-HBs are mixed with the test serum; any HBsAg present will attach to the antibody and agglutinate the cells within 20 min. A **latex slide test** based on the same principle is now available that can be read in 5 min.

A positive rapid test for HBsAg must always be confirmed by ELISA. The finding of HBsAg does not in itself mean that the patient is infective for others, although he or she should be treated with appropriate precautions until the result of a test for HBeAg is available. This takes longer than the rapid test, for it too must be done by ELISA. Some laboratories also test for DNA polymerase as a measure of virus replication. The full marker profile is completed by testing, again by ELISA, for the three antibodies, anti-HBs, anti-HBe, and anti-HBc.

Electron microscopy can be used to test quickly both for HBsAg and infective Dane particles, since they are so numerous when present that finding them in serum is easy; but this method is unsuitable for large numbers of specimens and is not in routine use.

Non-specific tests

These are useful for assessing liver function but do not distinguish between the various forms of hepatitis. In the acute phase or in exacerbations of chronic aggressive hepatitis the levels of alanine aminotransferase (ALT) and other liver enzymes are raised and prothrombin is depressed. Serum bilirubin is also increased.

3.7 Control measures

Person-to-person transmission

Horizontal spread of infection is blocked by preventing blood or body fluids from an infected person gaining access to the circulation of someone else. Degrees of infectivity can be assessed from the marker profile (Table 20.4).

Table 20.4 Infectivity of HBV carriers

Degree of infectivity	Markers				
	Antigens		Antibodies		
	HBsAg	HBeAg	Anti-HBs	Anti-HBe	Anti-HBc
High	+	+	−	−	+
Intermediate	+	−	−	+	+
Low	+	−	+	+	+

Health care staff must take the obvious personal precautions, such as keeping cuts and abrasions covered and wearing gloves when injecting or operating upon actual and potential high-risk patients. All hospitals should have detailed codes of practice for use in wards, theatres, clinics, and laboratories and it goes without saying that these must be meticulously observed. They include instructions for the use, whenever possible, of disposable equipment; sterilization by heat; chemical disinfection with hypochlorites or glutaraldehyde; and for the action to be taken in the event of spillages or injury of staff. (See Appendix A for further information). Special care is taken to exclude carriers from entering renal dialysis units, where, because of the depressed state of immunity of the patients, HBV infection readily becomes established with risks both to patients and staff. Stringent control measures have eliminated hepatitis B from dialysis units in the UK but it is still prevalent in some other countries.

Active immunization

The control of hepatitis B is being revolutionized by the use of vaccines prepared from HBsAg, which are both safe and effective.

The 'first-generation' vaccines were prepared from the blood plasma of carriers. This method was adopted because of inability to propagate HBV in quantity in the laboratory. It included elaborate purification and safety procedures to ensure the elimination of infective HBV and other viruses. These hazards have now been eliminated by the introduction of genetically engineered vaccines. The gene that codes for HBsAg has been cloned into yeast cells, which are cultured on a mass scale and produce large quantities of the antigen. The dosage of the vaccine employed in the UK is 20 µg of HBsAg given intramuscularly at 0, 1, and 6 months, with boosters at 5-year intervals for those at special risk. The third dose is essential to secure full protection. Some people, especially those aged over 40 years, fail to respond in terms of forming anti-HBs.

Although hepatitis B vaccines are in general highly effective, there have been reports of mutant strains against which they do not protect. In one such strain, the mutation affected only one amino acid in the HBsAg, a situation reminiscent of that with influenza (Chapter 10). It seems likely that such mutants are rare, but, clearly, a careful watch for them must be maintained. New approaches to the preparation of hepatitis B vaccines are discussed in Chapter 28.

In the developed countries, the candidates for vaccination are primarily the high-risk groups listed in Table 20.3. In areas with high prevalences of carriers, the first priority is immunization of newborn infants, with the ultimate aim of reducing the incidence of liver cancer.

Passive immunization

Human immunoglobulin with a high titre of anti-HBs (**HBIG**) can be used to provide immediate passive protection:

◆ with vaccine, for infants born to carrier mothers
◆ for health care staff who suffer 'needle-stick' or other penetrating injuries while attending actual or suspected high-risk patients. Adults need two or three doses at monthly intervals. In one well-conducted trial, a two-dose regime had a protective efficacy of 76 per cent.

3.8 Treatment

The acute infection does not normally need treatment, but the threat of an HBeAg-positive carrier state demands action. Some at least of those who become carriers in adult life are naturally deficient in interferon, and some cures have been obtained with large doses of **alpha-IFN**. Unfortunately, this treatment is of no use in those who become infected in infancy.

Various DNA polymerase inhibitors have also been tried, notably vidarabine, acyclovir, and foscarnet (Chapter 29), but it is too soon to say whether the effects of these drugs are more than temporary.

3.9 *Delta agent*

This curious little virus, sometimes known as hepatitis D, is 36 nm in diameter and contains ssRNA with a molecular weight of only 0.6×10^6. This is much too small to code for the usual virus functions and **its replication depends on that of HBV itself**. Delta agent is thus an incomplete virus, reminiscent of the parvoviruses described in Chapter 12. Its outer coat is in fact formed from HBsAg, the specific delta antigen itself being within the core.

This agent is transmitted along with HBV, either at the time of first infection with the latter (co-infection) or during a subsequent exposure (super-infection). Its effect is to exacerbate the hepatitis B infection, both in the acute phase and during relapses of chronic aggressive hepatitis. It is especially prevalent in Italy (where it was first discovered), parts of the Middle East, and Latin America. **Intravenous drug abusers** are particularly liable to acquire delta agent, and epidemics among them have been documented in Scandinavia and elsewhere.

Diagnosis of the active infection depends on detecting the delta antigen or its IgM antibody in the blood.

It is noteworthy that immunization against hepatitis B also protects against infection with delta agent.

4 NON-A, NON-B HEPATITIS

The development of methods for diagnosing hepatitis A and B revealed a residuum of cases with clinical and epidemiological features in common with these infections, but lacking the diagnostic markers. These infections—for there are more than one variety—are caused by viruses whose nature is as yet unclear. They are often referred to collectively as non-A, non-B (or NANB) viruses, but recent researches have started to provide these rather shadowy agents with individual identities.

4.1 *Hepatitis C*

During the 1970s, a form of hepatitis was recognized that had features in common with hepatitis B, but which was seronegative for HBV markers. It occurred after blood transfusion or injection of blood products, had an incubation period of about 60 days, and was transmissible to chimpanzees. It also seemed even more liable than HBV to cause severe liver damage. In a remarkable use of recombinant DNA techniques, involving the examination of more than a million clones for specific activity, a virus, now termed hepatitis C (HCV), was identified by Choo, Houghton, and co-workers in the USA. The agent is probably 60–70 nm in diameter and enveloped; it has an RNA genome with affinities to those of the flaviviruses. High incidences of antibodies to it have been found in patients with post-transfusion hepatitis, haemophiliacs receiving anti-clotting factors, intravenous drug abusers, and people with cirrhosis of the liver or hepatocellular carcinoma. In the UK, preliminary evidence suggests that the prevalence of anti-HCV in healthy blood donors is 0.5–1 per cent. In summary, therefore, HCV, although very different from HBV, is transmitted in a similar fashion; it may well prove more dangerous in terms of its ability to provoke liver cirrhosis and carcinoma. With the rapid development of diagnostic tests, its capacity for perinatal transmission will doubtless be determined in the near future.

4.2 Hepatitis E

As we work our way through the alphabet ('D' being used for the delta agent), hepatitis E seems an appropriate designation for the last agent on our list, since it (or they, for there may be more than one) is spread by the **enteric** route. It has a relatively short incubation period of 2–7 weeks and seems to be spread mainly in water. There have been large water-borne epidemics, notably in India. The virus is not well-characterized, but appears to be about 27 nm in diameter, and resembles calicivirus (Chapter 11). Both morbidity and mortality are worst in pregnant women, in whom the death rate may approach 40 per cent.

5 REMINDERS

◆ Several viruses cause hepatitis as part of a general infection, but the primary target cells of the 'true' hepatitis viruses are the hepatocytes. Except for HAV, none has been propagated in quantity in the laboratory but all are infective for chimpanzees.

◆ **Hepatitis A** is an **enterovirus** transmitted by the faecal-oral route. The incubation period is 2–6 weeks. Jaundice lasts about a month, after which recovery is complete in the great majority of cases. There is no carrier state.

◆ Laboratory diagnosis is made by finding **specific IgM antibody** in the serum or by **immune electron microscopy of faeces**.

◆ **Normal human immunoglobulin** confers temporary passive immunity and is useful for travellers going to highly endemic areas.

◆ **Hepatitis B** virus belongs to the **Hepadnaviridae**. It is a **DNA** virus with a circular genome and makes use of reverse transcription to replicate progeny DNA from an RNA intermediate.

◆ Transmission is by **blood or body fluids**, very small amounts of which may be infective for others. Hepatitis B is readily transmitted by **sexual intercourse**. The incubation period is usually 2–3 months (range 6–20 weeks).

◆ During active infection the blood contains numerous infective virions (Dane particles) and smaller spheres and tubules of surface antigen (**HBsAg**). The presence of **HBeAg** indicates that the blood contains Dane particles and is infective for others. The core antigen, HBcAg, is not detectable in the blood.

◆ The antibodies, **anti-HBc**, **anti-HBe**, and **anti-HBs**, appear in this order and indicate recovery. People who fail to form anti-HBe or anti-HBs become chronic carriers.

◆ HBe-positive carriers are at risk of **cirrhosis of the liver** and **hepatocellular carcinoma**. Babies acquiring the infection at birth nearly all come into this category. The high carrier rate in some countries, e.g. Africa, is correlated with a high incidence of liver cancer.

◆ **Treatment** of chronic hepatitis acquired in adulthood with **alpha interferon** is giving promising results.

◆ Prevention of hepatitis B depends on blocking person-to-person transmission and, particularly for large-scale control, **immunization**.

◆ **Delta agent** is an incomplete RNA virus dependent on HBV for replication. It may be transmitted along with hepatitis B, and exacerbates its course. It is becoming prevalent in intravenous drug abusers.

◆ There are at least two non-A, non-B hepatitis agents, **HCV and HEV**, transmitted respectively like HAV and HBV and with similar incubation periods. Diagnostic tests are becoming available for HCV, which may be a flavivirus. HCV is even more likely than HBV to cause chronic liver damage and cirrhosis.

PAPOVAVIRUSES

1 INTRODUCTION

The name of this family is yet another example of the acronyms so beloved by virologists; although Papova sounds like a Russian ballerina, its component syllables in fact help you to remember the characteristics of the viruses concerned.

◆ **pa** = papillomaviruses

◆ **po** = polyomaviruses

◆ **va** = vacuolating agents

The -oma components of these names immediately alert you to the fact that these agents have something to do with tumours and, indeed, the papillomaviruses have now displaced herpes simplex as the centre of attention in relation to cancers of the genital tract.

2 PROPERTIES OF THE VIRUSES

2.1 *Classification*

The family *Papoviridae* consists of two genera, the papillomaviruses and the polyomaviruses; the vacuolating agents belong to the latter genus, but are distinguished from the other members by their lack of association with any known disease.

2.2 *Morphology*

The virions are **icosahedral** and have no envelopes (Fig. 21.1). The papillomaviruses are about 55 nm in diameter and the polyomaviruses are slightly smaller, at around 45 nm.

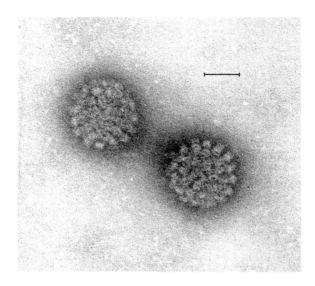

Fig. 21.1 Electron micrograph of papillomaviruses. Scale bar, 50 nm.

2.3 *Genomes*

The genomes are **circular dsDNA** and code for a comparatively small number of polypeptides, some of which are, however, important in transforming normal cells to the malignant state.

2.4 *Replication*

Replication of papovavirus DNA takes place within the host-cell **nucleus**. The papillomaviruses are something of a curiosity in that they can multiply only in **differentiating stratified squamous epithelium**, which cannot be grown as conventional cell cultures. Study of these agents by the usual methods has therefore been considerably hampered; but this difficulty has a bonus, because it pushed virologists into exploiting DNA technology for novel ways of detecting and classifying them. This concentration upon the characteristics and functioning of the genome has taught us much about the mechanisms of oncogenesis. The polyomaviruses, on the other hand, can readily be propagated in cell cultures.

3 CLINICAL AND PATHOLOGICAL ASPECTS

3.1 *Diseases caused by papillomaviruses*

Papillomaviruses or their antigens have been detected in a wide range of mammals and in some birds. Because, for the reason just explained, they cannot be isolated and distinguished by serological tests, they are **classified on the basis of their degree of DNA homology**, i.e. on how closely their nucleotide sequences correspond. About 60 types of human papillomavirus (HPV) have now been identified; they cause disease only in skin and mucous membranes, where they give rise to warty lesions. These are usually benign, but some may become malignant; there are distinct — but not invariable — associations between the type of HPV, the anatomical site involved, and the potential to cause malignant lesions. Table 21.1 gives examples of lesions caused by papillomaviruses and the types of virus most commonly associated with them; this list is not exhaustive.

Table 21.1 Lesions caused by human papillomaviruses (HPV)

Type of lesion	Site	Predominant HPV types*
Predominantly benign lesions		
Common warts	Skin, various sites	2, 4
Plantar and palmar warts	Hands, feet	1, 2, 4
Butchers' warts	Hands	7
Flat warts	Skin, various sites	3
Genital warts (condylomata acuminata)	Cervix, vulva, penis	6, 11
Juvenile laryngeal papilloma	Larynx	6, 11
Malignant or potentially malignant lesions		
Flat warts	Skin	**10**
Bowenoid papulosis	Vulva, penis	**16**
Premalignant intraepithelial neoplasia	Cervix, penis	6, 11, **16**, **18**, 31, 40, 42, 44
Carcinoma	Cervix, penis	**16**, **18**
Papilloma/carcinoma	Larynx	**16**
Epidermodysplasia verruciformis	Skin, various sites	Many types, including **5**, **8**, 9, **10**, 12, **14**, 15, 17, 19–29

* HPV types particularly liable to cause malignancies are printed in **bold**.

Predominantly benign lesions

Cutaneous warts. The virus is transmitted from infected skin, either by direct contact or through fomites, and enters its new host through abrasions. Swimming pools and changing rooms are fertile sources of infection, and skin warts are most liable to affect the younger age groups.

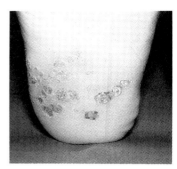

Fig. 21.2 Plantar warts. (Courtesy of Department of Dermatology, The Royal London Hospital, London.)

The **common wart (verruca vulgaris)** has a characteristically roughened surface; the excrescences are usually a few millimetres in diameter and may occur in quite large numbers anywhere on the skin, but especially on the hands, knees, and feet (Fig. 21.2). They usually cause no inconvenience, apart from the cosmetic aspect, but plantar warts on the soles of the feet are pressed inward by the weight of the body; they may be painful and call for more urgent treatment than those in other sites. **Flat warts (verrucae planae)**, as their name implies, are flatter and smoother than common warts; they also predominantly affect children. **'Butchers' warts'**, caused by HPV-7, are an occupational hazard. Why only this type affects butchers—and only butchers—is a mystery; there is certainly no relationship between it and any of the bovine papillomaviruses (BPV).

Genital warts. It is at this point, as we work down through Table 21.1, that we arrive at the somewhat blurred interface between the benign and malignant lesions caused by papillomaviruses.

As might be expected, genital warts are acquired by sexual contact: they are, in fact, one of the most common sexually transmitted diseases, and often occur in association with others, e.g. gonorrhoea or chlamydial infection. It follows that their highest incidence is in the late teens and early adulthood.

Ordinary skin warts sometimes occur on the penis, but the most frequently seen lesions, both in men and women are **condylomata acuminata** (condyloma acuminatum, if, as rarely happens, there is only one of them). This Greco-Latin name means 'pointed lump', but sounds nicer. They are fleshy, moist, and vascular and may grow much larger than the common skin warts. By contrast with the latter, they may be pointed or filiform.

Condylomata acuminata may be seen at the following sites.

In men.

◆ On the **penis**, affecting the area around the glans and prepuce more than the shaft (Fig. 21.3(a)).

◆ Within the **urethral meatus** and **urethra itself**.

◆ Around the **anus** and within the **rectum**, particularly in homosexuals; in those practising receptive anal intercourse this site is more frequently affected than the penis.

It is unusual for condylomata acuminata to become malignant, but they occasionally progress to squamous cell carcinoma. This is particularly true of the comparatively rare giant condyloma, which forms huge masses of tissue on the penis or in the anoperineal area.

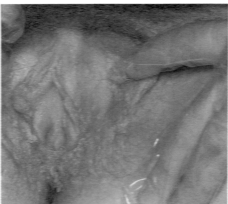
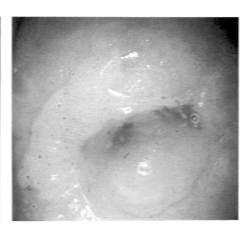

In women

◆ On the **vulva** (Fig. 21.3(b)).

◆ Occasionally, in the vagina.

◆ On the **cervix**. Here, the typical lesion is the flat, intraepithelial type (Fig. 21.3(c)), rather than the fleshy variety seen on the external genitalia. It is difficult to distinguish clinically between this lesion, caused by a papillomavirus, and other forms of cervical dysplasia.

◆ Around the **anus** and on the **perineum**.

In children. These lesions have been observed on the external genitalia of both male and female infants and children. Although they may be indicators of sexual abuse, it is dangerous to jump to this conclusion, since the infection may be transmitted from the mother during delivery, or even by close but innocent contact within the family.

Fig. 21.3 Genital warts. **(a)** Penile warts (condylomata acuminata). (Reproduced with permission from Robinson, T.W.E. and Heath, R.B. (1983). Virus diseases and the skin. Churchill Livingstone, Edinburgh.) **(b)** Vulval warts (condylomata acuminata). **(c)** Flat condylomata of cervix. (Parts **(b)** and **(c)** by courtesy of Dr David Oriel.)

Malignant or potentially malignant lesions

Bowenoid papulosis. This syndrome manifests as multiple papules on the penis or vulva; it is usually seen in young people, and, although usually benign, may become malignant. It takes its name from Bowen's disease, which occurs in older people and is not associated with HPV, but with which, however, there are histological resemblances.

Premalignant intra-epithelial dysplasia. Irregularities in the histological pattern of the epithelium ('**atypia**') may occur on the penis, vulva (vulvar intraepithelial neoplasia, VIN), vagina (VAIN), or cervix (CIN). They are staged according to the degree of dysplasia and the most severe form (CIN 3 in the case of cervical lesions), which involves all layers of the stratified epithelium, has a high chance of progression to metastasizing carcinoma. It is now known that many of these lesions are associated with papillomaviruses. **HPV types 16** and **18**, in particular, are heavily implicated in the causation of **carcinoma** both of the **cervix** and of the **penis**.

Laryngeal infections with HPV. It is probably no accident that the types of HPV causing warts in the **juvenile larynx** are the same as those associated with genital warts, namely **HPV-6** and **11**; they may well be transmitted during delivery and establish a persistent infection. Although they tend to recur after treatment, these papillomas do not become malignant in children; this is not however true of the **adult** variety, associated with **HPV-16**, which may develop carcinomatous changes.

Epidermodysplasia verruciformis. This is a rare autosomal recessive disease associated with defects in T cell function and numbers, but not with other varieties of immune deficiency. The multiple flat lesions may persist for years; they are associated with a wide range of HPV types and may become malignant, especially in areas of skin exposed to sunlight.

Immunosuppressed patients. Allograft recipients receiving immunosuppressive treatment are liable to develop squamous cell carcinomas of the skin; HPV-5 has been demonstrated in the lesions of at least one such patient.

Pathology

A characteristic feature of the benign skin warts is **hyperkeratosis**, i.e. a massive proliferation of the keratinized layers of the dermis (Fig. 21.4). Particularly in plantar warts, there are deep extensions of the hypertrophic epidermis. Large, pale, vacuolated cells are present in the granular layer; these are more pronounced in plane warts than in verruca vulgaris. There are many eosinophilic cytoplasmic inclusions which are not of viral origin, but which consist

Fig. 21.4 Section of skin wart. Phloxine–tartrazine stain. The epidermis is hyperplastic. The epidermal cells, particularly those at the left in the body of the wart, contain many darkly stained inclusions.

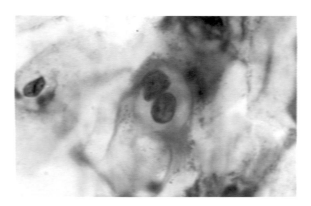

Fig. 21.5 Koilocyte in cervical scraping. The double nucleus, which is characteristic, is surrounded by an area of unstained cytoplasm. (Courtesy of Dr David Oriel.)

of abnormally large keratohyalin granules. There are however basophilic inclusions in the nuclei of the epidermal cells in which typical papillomavirus virions can be seen by electron microscopy.

The histology of condylomata acuminata is quite similar, except that hyperkeratosis is not a significant feature. The presence of vacuolated cells in cervical scrapings (Fig. 21.5) is diagnostically useful; they are known as koilocytes ('empty cells') because of their appearance in stained preparations, but the term is inappropriate since the apparently empty vacuoles contain enormous numbers of virions.

At this point, the histopathology of papillomavirus lesions merges into that of precancerous and cancerous changes, which is beyond the scope of this book.

Immune response

The failure to culture papillomaviruses has greatly hindered our ability to unravel the immune reactions to infection. This is a pity, because the little that we do know poses tantalizing suggestions that the immunology of papillomavirus infections may have some unique characteristics. Why, for example, may cutaneous warts persist for years and then suddenly disappear for no obvious reason? What is the relationship between the immune response and oncogenesis? What are the respective roles of cell-mediated immunity (CMI) and antibody? We know from immunofluorescence studies that both IgM and IgG antibodies may appear, particularly in those with regressing warts, but not all those with antibody have warts and not all warty individuals have antibody. On the other hand, CMI may be equally or more important, since regressing lesions contain many lymphocytes, predominantly of the T-suppressor (CD8) subclass, but also T-helper (CD4) cells. Circumstantial evidence of the importance of the immune responses is provided by patients with epidermodysplasia verruciformis, who have an inborn defect of immunity, and by the high prevalence of warts in immunosuppressed patients.

The mechanisms by which papilloma and other viruses give rise to malignant tumours is dealt with in Chapter 6.

Laboratory diagnosis

The presence of warty lesions is usually so obvious to the eye, aided when necessary by a magnifying glass, colposcope, or urethroscope, that laboratory methods are largely redundant except for determining, if possible, whether a given lesion is benign or malignant. The histological appearances are characteristic, and Papanicolau staining of smears may reveal diagnostic features, notably the presence of vacuolated cells (**koilocytes**). The presence of HPV — but not its type — can be verified by an antiserum against disrupted BPV particles. For typing HPVs within lesions, recourse must be had to DNA hybridization techniques, which are not as yet routine procedures.

Treatment

Skin warts may need treatment on cosmetic grounds, or, in the case of plantar warts, because of pain and disability on walking. They may be removed by treatment with **podophyllin**, extracted from the roots of the American mandrake, or by **freezing** with liquid nitrogen. It is a curious but quite well authenticated observation that common skin warts can sometimes be made to disappear by suggestion. There are several methods of dealing with cervical lesions, including **freezing**, **electrodiathermy**, and **cone biopsy**. Recurrent laryngeal warts may be removed surgically, but must never be treated with irradiation, as this induces malignant changes. Injection of **interferon** directly into the lesions has met with limited success, but by no means in all patients so treated.

3.2   *Diseases caused by polyomaviruses*

Poly-oma means what it sounds like — 'many tumours'. In 1951, Ludwik Gross, in the USA, injected tissues from mice suffering from a naturally occurring leukaemia into newborn mice; unexpectedly, they did not develop leukaemia but malignant tumours of the salivary glands (see Chapter 6). After a number of passages in the laboratory, the agent was able to cause various types of tumour, which accounts for its name. This property was very noticeable when it was injected into other species of animal. Polyomavirus naturally infects wild mice, being transmitted from mothers to offspring. Under these conditions the virus does not cause tumours. Instead, a persistent, life-long infection is established, but without overt disease. (The leukaemia affecting the mice from which polyomavirus was first isolated proved to be due to concurrent infection with a different agent, a retrovirus.)

Here, then, we have other papovaviruses with oncogenic potential, but in terms of their ability to cause disease in humans they bear no resemblance whatsoever to HPV. Another important difference is their ability to grow and induce cytopathic effects in cultured cells, notably those from embryo mice.

JC virus and progressive multifocal leucoencephalopathy (PML)

The year 1958 saw the first description of this syndrome, which is characterized by hemiparesis, disturbances of speech and vision, dementia, and a relentless progression to death within a few months. It is noteworthy that PML, which is quite rare, often occurs in **elderly** patients already suffering from disease of the **reticuloendothelial system**; the first three to be described suffered either from chronic lymphatic leukaemia or from Hodgkin's disease. The lesions in the brain comprise multiple areas of **demyelination**, **abnormal oligodendrocytes**, and, in the later stages, pronounced **astrocytosis**. The name of the syndrome derives from these clinical and pathological features.

A few years later, virions with the morphology of polyomavirus were found in the nuclei of affected glial cells, and, in 1970, a virus, termed JC after the initials of the patient, was isolated from PML in cultures of human fetal glial cells. This virus is widespread in humans, in whom it causes persistent infection, primarily in the urinary tract, until stirred into more dangerous activity by immunosuppression. PML is almost unheard of in immunologically normal people.

PML was a comparatively rare disease until, quite recently, it came to greater prominence as a presenting feature or complication of AIDS, which is not surprising in these severely immunocompromised patients.

BK virus: another persistent agent

A year after the isolation of JC virus, Sylvia Gardner and her colleagues isolated a similar agent, termed BK, from the urine of an immunosuppressed renal transplant patient in London. This virus grows readily in monkey kidney (VERO) cells and has now been isolated from the urine of many immunocompromised patients. Although its physical characteristics are very similar to those of JC virus, its pathogenicity for humans is quite different.

As determined by tests for antibody, the prevalence in the general population of BK is also high; like wild mice, people seem to become infected with polyomaviruses early in life without suffering obvious ill effects. Normal people do not excrete BK virus unless they are immunocompromised or pregnant, in both of which cases the virus is liable to reactivate. Even when it does, however, overt disease is rare, the main exception being small numbers of cases of ureteric stenosis in renal transplant recipients and of acute haemorrhagic urethritis. BK virus does not appear to cause PML.

Vacuolating agents

During the 1950s, the widespread use of monkey kidney cells for isolating polioviruses and preparing vaccine brought to light many viruses that caused inapparent infections; not until the cells were explanted into culture vessels did cytopathic effects become obvious. These agents were termed 'simian vacuolating' viruses (SV). One of them, which induced a vacuolated appearance in the cytoplasm, was termed SV40. This agent proved to be a papovavirus, and like other members of the genus was oncogenic, this time for hamsters. It did not, however, seem to cause disease in the monkeys from which it was isolated, and it would probably have remained of academic interest but for the fact that it proved comparatively resistant to formalin, which was used to inactivate the poliovaccine widely used at that time. Unfortunately, many thousands of doses had been issued before the discovery of SV40 and there was a real worry that some of the inoculated children would develop tumours in later life. Luckily, intensive follow-ups have revealed no evidence of any such effects, but the scare created by this episode did much to improve the testing of biological products to ensure that they contain no extraneous viruses.

Immune response

By contrast with papilloma viruses, polyoma agents induce a good **antibody response**, which is measurable by a haemagglutination-inhibition test employing human group O erythrocytes. Little is known about cell-mediated immunity. In view of the persistent nature of these infections, it seems that immune mechanisms serve only to keep the viruses at bay, without actually eradicating them.

Laboratory diagnosis

The clinical diagnosis of PML can be confirmed at autopsy by the characteristic histological appearances. Isolation of JC virus is difficult and can be done only in a very few specialized laboratories. Papovavirus virions, often in crystalline arrays, can be found by electron microscopy in ultrathin sections of oligodendrocyte nuclei.

Isolation of BK infections by cultural methods is slow; a faster diagnosis may be made by staining exfoliated cells in the urine for viral antigen by immunofluorescence or immunoperoxidase methods.

Treatment

Limited success has been claimed for cytarabine in the treatment of PML, but for practical purposes the prognosis is dire. By and large there is no indication for treating BK infections, which is fortunate since there is no specific therapy.

 # REMINDERS

◆ The papovaviruses are non-enveloped and icosahedral; the genome is **dsDNA**.

◆ The two genera are ***Papillomavirus*** and ***Polyomavirus***.

◆ **Human papillomaviruses (HPV)** replicate only in **differentiating squamous epithelium** and thus cannot be grown in conventional cell cultures.

◆ HPV cause **warts** in **skin** and **mucous membranes**, which are usually benign but may become malignant, especially in the genital tract. Certain types of HPV, notably **16** and **18**, are strongly implicated in the causation of cancer of the genital tract, notably **carcinoma of the cervix**.

◆ HPV are particularly liable to cause malignant disease in **immonocompromised patients**.

◆ **Polyomaviruses** are widespread in humans and animals, causing **persistent but silent infections of the urinary tract**. They are, however, oncogenic if injected into newborn animals or into other species.

◆ The two best characterized human polyomaviruses are JC and BK.

◆ **JC virus** causes **progressive multifocal leucoencephalopathy (PML)**, a rare fatal demyelinating disease of the CNS, usually seen in later life and particularly affecting those with disease of the reticuloendothelial system.

◆ **BK virus** also reactivates in **immunocompromised patients** but rarely causes overt disease.

◆ Another polyomavirus, SV40, causes persistent but inapparent infection of the kidneys of some monkeys, and is liable to be present in cultures prepared from them.

RETROVIRUSES AND AIDS

1 INTRODUCTION

The first discovery of a retrovirus was made as long ago as 1910 by Peyton Rous working at the Rockefeller Institute for Medical Research in New York, and this virus, avian sarcoma virus, induced tumours in muscle, bone, and other tissues of chickens (Chapter 6). It was not until the 1930s that further retroviruses, causing tumours in mice and other mammals, were discovered. But these viruses were still regarded as laboratory curiosities until Jarrett, in Glasgow, described feline leukaemia virus, which appeared to spread naturally in household cats. Perhaps as a portent for the later discovery of HIV-1 in humans, it caused an **immune deficiency** in infected animals. These viruses possess a unique enzyme, reverse transcriptase (RT), which uses the viral RNA as a template for making a DNA copy, which then integrates into the chromosome of the host cell, and there serves either as a basis for viral replication or as an oncogene (see Chapter 6). The discovery of RT overturned a central dogma of molecular biology — that genetic information flows in one direction only, from DNA to messenger RNA to protein.

1.1 *The discovery of the human retroviruses*

In 1978, Robert Gallo, in the USA, isolated a retrovirus from the lymphocytes of a leukaemia patient that had been maintained in culture by a new technique involving stimulation of the cells with interleukin-2. Japanese workers had earlier noticed clustering of cases of adult T cell leukaemia-lymphoma in the southern islands of Japan. To epidemiologists this clustering hinted at an infectious aetiology. A retrovirus was subsequently isolated from a sufferer and nucleotide sequence analysis of the viral genome showed it to be identical to Dr Gallo's isolate. The first human retrovirus to be discovered was named **human T cell leukaemia virus type I (HTLV-I)**. Epidemiological studies showed that HTLV-I was endemic not only in Southern Japan but also in parts of the USA, the Caribbean, and Africa. It has been suggested that HTLV-I originated in Africa, where it infected Old World primates; after transmission to humans, it may have reached the Americas along with the slave trade, and then the southern islands of Japan with early Portuguese explorers who had previously been in Africa.

A second virus, HTLV-II, was isolated in Seattle, USA, from the cells of a patient with a rare 'hairy cell' leukaemia, but little is known about this virus at present.

In 1981 a new clinical syndrome, characterized by profound immuno-deficiency, was recorded in male homosexuals and was termed the acquired immunodeficiency syndrome (AIDS). The Centers for Disease Control in Atlanta reported an unusual prevalence of *Pneumocystis carinii* pneumonia (PCP) in a group of young, previously healthy, male homosexuals. Before then, this parasite had been associated with disease only in patients whose immune systems had been seriously impaired as a result of drug therapy or congenital cellular immune deficiency. At the same time came reports of previously healthy young homosexuals in New York and San Francisco who had developed a rare cancer, Kaposi's sarcoma. The first isolation of a retrovirus from an AIDS case was made by Luc Montagnier and Barré-Sinoussi at the Pasteur Institute in Paris early in 1983 and quickly confirmed by Robert Gallo. The first case of AIDS in the UK was diagnosed in late 1981 in a homosexual from Bournemouth, after he returned home from Miami. An AIDS research trust was established in his name, Terrence Higgins.

A fourth human retrovirus, HIV-2, was isolated from mildly immuno-suppressed patients in West Africa.

2 PROPERTIES OF THE VIRUSES

2.1 *Classification*

The causative agent of AIDS, namely HIV-1, is a member of the *Retroviridae* by virtue of its RNA genome, morphology, and possession of RT. Particular characteristics, such as the absence of oncogenicity, the lengthy and insidious onset of clinical signs, and its distinctive morphology at the virus budding stage, categorize the agent as human lentivirus (Latin *lentus*: slow). Human lentiviruses have a unique genome with four regulatory genes (see Section 2.3) although they share other genes such as *pol*, *env*, and *gag* with the other

Table 22.1 Classification of retroviruses infecting primates

Subfamily	Disease caused	Natural hosts
Oncovirinae		
Human T-cell leukaemia virus HTLV-I	Adult T-cell leukaemia, lymphoma, tropical spastic paraparesis	Humans
Human T-cell leukaemia virus HTLV-II	Hairy cell leukaemia	Humans
Spumavirinae	Inapparent persistent infections	Primates and other animals
Lentivirinae		
Human immunodeficiency virus (HIV-1)	Immune deficiency, encephalopathy. Virus can infect chimpanzees but causes no clinical signs	Humans
Human immunodeficiency virus (HIV-2)	Immune deficiency. Less pathogenic than HIV-1	Humans and primates
Simian immunodeficiency virus (SIV-I)	Immune deficiency. No disease in wild African green monkeys but AIDS in rhesus monkeys	Monkeys

retroviruses. (In accordance with convention, the names of genes are italicized, whereas those of the polypeptides that they encode — the gene products — are not.) Another important lentivirus is HIV-2, first isolated in Africa, where it has caused extensive infection but comparatively mild disease.

The other human retroviruses, HTLV-I and II, are classified in a different subfamily because of their different genome structures and their ability to cause tumours rather than immunosuppression.

Table 22.1 lists the members of the *Retroviridae* and their diseases. Note that roman and arabic numerals are used to designate strains of HTLV and HIV/SIV respectively.

2.2 *Morphology*

The typical retrovirus particle is 100 nm in diameter with an outer envelope of lipid penetrated by glycoprotein spikes (Fig. 22.1), the **envelope (env) protein**. In the case of HIV-1 the lipid envelope encloses an **icosahedral-**shaped shell of protein within which is a vase-shaped protein core containing two molecules of RNA in the form of a ribonucleoprotein. Bound to the diploid RNA genome are several copies of the RT enzyme (see also Fig. 1.7).

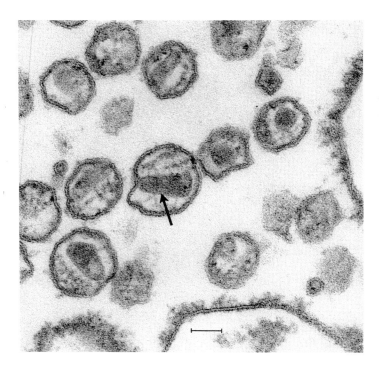

Fig. 22.1 Electron micrograph of HIV-1. vase-shaped core of the virion is arrowed. (Courtesy of Dr David Hockley.) Scale bar, 50 nm.

2.3 | *Genome and polypeptides*

The organization of the **ssRNA** genome of HIV-1, **approximately 10 000 nucleotides** in length, is complex. Unlike certain oncogenic viruses in the family, it has no *onc* gene, but has some unique features, particularly the possession of three additional control genes which can enhance viral replication (*rev*, *tat*, and *vif*) and a repressor gene, *nef*.

At least 15 viral polypeptides are detectable in cells infected with HIV and many of these have structural functions in the virus particle itself. The two most important viral proteins, particularly from the point of view of laboratory diagnosis, are the env and gag (group antigen) polypeptides. The env polypeptide is composed of two subunits, the outer glycoprotein knob (gp120) and a transmembrane portion (gp41) which joins the knob to the envelope of the virus. Three gag-related polypeptides are cleaved from a single **polyprotein**, one of which, p24, is extensively used in diagnostic testing.

The *pol* gene codes for at least three proteins, the largest being RT; another is an integrase enzyme with the important function of integrating the HIV proviral genome into cellular DNA.

There is considerable genetic variation between different isolates of HIV-1 virus. In the case of the env protein, the degree of variation is comparable with that between subtypes of influenza A virus, amino-acid sequences differing by up to 25 per cent.

2.4 *Replication*

HIV-1 binds specifically to the CD4 receptor which is expressed on the surface of certain T lymphocytes — the T-helper cells (Chapter 5). But, since it also infects B lymphocytes, macrophages, dendritic cells, and brain cells, these too may have a few CD4 receptors.

The virus penetrates the cell by fusion from without (Chapter 2). The RT enzyme is able to transcribe the viral RNA and make a DNA copy. The RNA portion of the molecule is digested away and a dsDNA form synthesized. This circularizes but, just before integration as a proviral DNA, becomes a linear molecule again. Integration is necessary to establish the latent state but virus can replicate without integrating in cellular DNA. Many details of replication are as yet unclear, but it is apparent that it involves a remarkably complex system of gene regulation of transcription. It is likely that the first viral mRNAs expressed in the infected cell are those coding for *rev* and *tat* gene products. The tat protein acts as a master switch at the beginning of the long terminal repeat (LTR) portion of the genome and speeds up the expression of other viral genes. The rev protein, itself under the control of tat, also stimulates the synthesis of other mRNAs coding for *gag*, *pol*, and *env* gene products. The gag and pol proteins are synthesized from full-length viral mRNAs and subsequently cleaved from a single large protein by proteolysis, whereas the env protein is synthesized from a subgenomic mRNA. Retroviruses, including HIV-1, are released from the infected cell by **budding** from the plasma membrane but even after release from the cell internal rearrangements of viral proteins take place, catalysed by a viral protease.

3 CLINICAL AND PATHOLOGICAL ASPECTS

3.1 *Clinical features of AIDS*

Adults

AIDS is a unique epidemic immunosuppressive viral disease that predisposes to life-threatening infections with 'opportunistic' organisms such as *P. carinii*, cytomegalovirus, atypical mycobacteria, *Toxoplasma gondii*, *Candida* spp., herpes simplex, and *Cryptosporidium*. Another highly distinctive feature of AIDS, particularly in homosexuals, is **Kaposi's' sarcoma**, known previously as a disease of the elderly but taking an aggressive form in AIDS patients, often involving multiple cutaneous sites and viscera. The exact cell of origin of this cancer is not known; it is multifocal (but not metastatic) with a predilection for the gastrointestinal tract and skin. In AIDS patients the skin lesions range from small pink spots to raised lesions of deep purple or blue, about 1 cm across. They are usually painless and appear anywhere on the skin or mucous membranes.

Table 22.2 Clinical stratification of HIV-related illness

Category*	Clinical features†
1	Asymptomatic
2	Immunothrombocytopenic purpura
3	Unexplained palpable lymphadenopathy of more than 4 months duration. Low-grade intermittent fevers (<38.5°C). Night sweats.
4	Minor opportunistic infection such as unexplained thrush or herpes zoster in individuals <60 years of age with or without lymphadenopathy
5	Intermittent or continuous fever (>38.5°C) for 1 month or watery diarrhoea for 2 weeks, or sustained weight loss
6 and 7	AIDS with or without Kaposi's sarcoma

* Stages 3, 4, and 5, together or separately are referred to as AIDS-related complex (ARC).
† The asymptomatic stage is often preceded by an acute but mild influenza-like illness, itself with an incubation period of 1 month.

The events following infection are illustrated in Table 22.2 which summarizes the generally accepted nomenclature scheme for the different stages of the disease. After a mild acute influenza-like illness with an incubation time of around 1 month, the disease becomes quiescent (stage 1). Approximately 60 per cent of asymptomatic cases move into the AIDS-related complex (ARC) stage of the disease (stages 3, 4, and 5) within the next 4 years. This is characterized by fever, weight loss, persistent lymphadenopathy, night sweats, and diarrhoea. Patients in this group proceed inexorably to full-blown AIDS (stages 6 and 7) commonly heralded by thrush, herpes zoster, and more seriously by *Pneumocystis carinii* pneumonia. The time from infection to death may be as long as 10 years.

It should be emphasized that much remains to be learned about the disease. It is already apparent that many AIDS patients have neurological signs such as dementia, the result of viral infection of the brain. As the survival time has increased, because of clinical intervention with the antiviral agent azidothymidine (AZT) and with the more efficient use of pentamidine to treat *P. carinii* pneumonia, the neurological sequelae have become more evident.

The signs and symptoms of AIDS may vary somewhat in the different categories of persons infected with the virus. Kaposi's sarcoma is less common, for example, in haemophilia patients, who have been infected as a result of infection of virus-contaminated factor VIII preparations.

Children

There is clear evidence of transplacental HIV infection; indeed, this is now recognized as the second most common mode of transmission. It is probable that transmission can also take place during delivery or from breast milk.

The clinical features of AIDS in a child are not dissimilar from adult AIDS but their delineation is often complicated by concomitant drug-related illnesses, since a proportion of such babies are born to drug-abusing mothers and are themselves drug-addicted. A child is considered to have ARC if interstitial pneumonitis, persistent candidiasis, or parotid swelling is present for 2 months together with two or more of the following: persistent general lymphadenopathy, recurrent bacterial infection, hepato- or splenomegaly, and chronic diarrhoea. The mortality of children with AIDS is high but the outcome in those with less severe disease or symptomless infection is less clear. The risk of paediatric AIDS is certainly higher in babies born to mothers who are symptomatic rather than simply seropositive.

Treatment

No cure exists for this retroviral infection and it is difficult to envisage the development of any antiviral drug which will both repress viral replication and also excise an integrated viral genome. The antiviral compound zidovudine (azidothymidine, AZT, Retrovir), has been shown to prolong the lifetime of sufferers by about a year (see Chapter 29). However, the compound is highly toxic for bone marrow cells; in the first clinical trials approximately one-third of treated persons required blood transfusions. Some patients are now treated with lower doses, and at an earlier stage of the disease. A chemically related dideoxynucleoside analogue, ddI, and foscarnet, a pyrophosphate analogue, also reduce viraemia in AIDS patients and are typical of the many compounds undergoing clinical trials.

Opportunistic bacterial, parasitic, and viral infections associated with AIDS may be held in check, at least for some time, by specific antimicrobial therapy, including acyclovir for herpes and varicella-zoster infections (Table 22.3).

Table 22.3 Treatment of HIV-1 and accompanying opportunistic infections

Infective agent	Drug
HIV-1	Zidovudine alone or with didanosine (ddI)
Herpes simplex	Acyclovir
Varicella-zoster	Acyclovir
Cytomegalovirus	Foscavir (retinitis), Ganciclovir (pneumonia)
Cryptosporidium	Amphotericin
Toxoplasma gondii	Sulphadiazine and pyrimethamine
Pneumocystic carinii	Trimethoprim, sulphamethoxazole, pentamidine
Mycobacteria	Streptomycin, isoniazid, rifampicin, p-aminosalicylic acid (PAS)

3.2 Clinical features of HTLV-I and HTLV-II infections

HTLV-I

HTLV-I frequently presents in adults as a lymphoma of the skin, lymph nodes, or both. Hepatic and splenic involvement are common and a further important diagnostic feature is hypercalcaemia, with or without bone lesions. The leukaemic phase of the disease is not always evident. Variants of what is usually an acute and aggressive disease occur and include a chronic T cell lymphocytosis. A form of 'smouldering' adult T cell leukaemia-lymphoma (ATLL) is also seen, in which the patients present with skin papules, nodules, and erythema for several years before the onset of leukaemia. Most seropositive patients do not develop overt disease and there are many asymptomatic carriers in endemic areas; the overall chance of ATLL developing in an HTLV-I seropositive person is estimated to be about 1 in 80.

In Martinique, West Indies, most patients with the chronic neurological disease, tropical spastic paraparesis (TSP), sometimes called HTLV-I-associated myelopathy (HAM), have HTLV-I antibodies. Overall, more than 75 per cent of reported TSP cases from the Caribbean area and Japan are HTLV-I antibody-positive.

HTLV-II

HTLV-II is known to cause a 'hairy (T) cell' leukaemia but very few cases have been described.

3.3 Pathogenesis

HIV-1 and HIV-2

HIV-1 enters the body via the bloodstream either during sexual intercourse, needle drug abuse, transfusion with contaminated blood products, or via the placenta (maternal–fetal transmission). There have been a number of instances of patients infecting health care staff or vice versa, e.g. during surgical or dental procedures. Virus-infected lymphocytes are present in sperm and may infect via microscopic breakages in the endothelial lining of the vagina or rectum. It is assumed that, initially, CD4 lymphocytes, macrophages, and dendritic cells are infected by the virus. But, since viral antigen has been found in only one in 10 000 lymphocytes, the reason for their ultimate dysfunction and death is a subject of some speculation. HIV-1 is strongly cytopathic and can cause cell-to-cell fusion resulting in the formation of giant syncytia and cell death. Viral glycoprotein present on the surface of an infected cell can interact with the CD4 receptors on many adjacent uninfected cells, thus multiplying the effect.

The replication of HIV genome is enhanced in antigen-stimulated T cells and it is assumed that persons with concomitant infections that stimulate T cell replication stand a greater chance of succumbing to AIDS. Macrophages are also infected with HIV and are themselves similarly stimulated by other antigens. They may act as a reservoir of the virus. As infection progresses, B lymphocyte functions are affected through their regulation by CD4-positive (T_h) cells (Chapter 5). The destruction of the CD4 helper T cell subset is particularly damaging to the overall orchestrated immune response of the host. This malfunction of the immune response leads to the appearance of opportunistic organisms which are normally held in check by immune T cells. HIV-1 may also reside and replicate in cells in the brain, causing cell destruction and neurological effects.

HTLV-I and HTLV-II

HTLV-I is transmitted by intravenous drug abuse, sexual intercourse, and blood transfusion.

It is intriguing that the pathological features of tropical spastic paraparesis (TSP) compare with those in the occasional patient dying with acute multiple sclerosis (MS), particularly with brainstem lesions. The histological picture may provide the clue as to how TSP and ATLL, two apparently dissimilar conditions, could be related to infection by the same virus. The prime suspect for initiating disease must be the HTLV-I-infected lymphocyte, transformed in ATLL and presumably chronically activated in TSP. Neuronal damage and demyelination are probably consequences of the inflammation, so that HTLV-I may not be truly neurotropic. Since only a very few cases of both diseases in the same patient have been reported, despite the many hundreds of each in Japan and the Caribbean, infected lymphocytes seem to be committed to produce one disease or the other or, much more commonly, neither.

Little is known about the pathogenesis of HTLV-II.

3.4   Immune response to HIV and HTLV

Figure 22.2 summarizes the rather complicated serological response of the host to HIV-1 infection. Antibodies to the env protein develop slowly and remain at high levels throughout the infection. Antibodies to the internal p24 protein have a different temporal pattern and rise during the early stages, only to decrease in parallel with the onset of serious signs of the disease. Presumably the host produces antibodies to all virus-induced proteins. An antibody-mediated cytotoxic response is also generated as well as a host-restricted cytotoxic T cell response to the structural env and gag proteins.

Much less is known about the dynamics of the immune response to the other human retroviruses. Antibodies to the respective env proteins can be clearly distinguished and hence are useful for differential diagnosis of the two lentiviruses and the T cell leukaemia viruses.

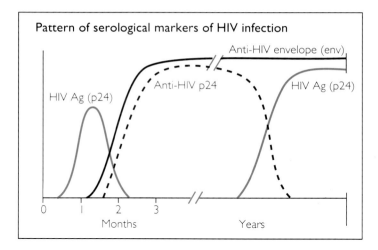

Fig. 22.2 The time-course of development of HIV antigens and antibodies.

3.5 *Epidemiology*

HIV-1 and HIV-2

The isolation of the causative virus of AIDS and the unfolding of the epidemic has caused a minor revolution in our moral and ethical values. It is likely that this viral pandemic will affect us all directly or indirectly for the next two decades or more. In 1990, there were 29 000 diagnosed cases of AIDS in Europe and 113 000 in the USA. As at December 1991 there were 5451 cases in the UK, of whom 288 were women and 78, children. The case fatality rate is about 60 per cent. The actual number of persons infected is a matter for conjecture but may reach over two million in the USA alone.

Analysis of the first 110 000 cases of AIDS in the USA (Table 22.4) has shown that high-risk groups are homosexual and bisexual men, intravenous drug users, recipients of contaminated blood or blood products, heterosexual contacts of persons in high-risk groups, and infants of high-risk parents. The age distribution of cases is illustrated in Fig. 22.3. WHO now describes AIDS as a typical venereal disease, but with a high mortality.

In the absence of effective vaccines or antivirals, prevention of infection can only be accomplished by individual action (Table 22.5) combined with screening of potentially contaminated blood and blood products to prevent iatrogenic infections. According to data reported by the World Health Organization, over 150 countries have cases of AIDS.

In Africa and Latin America the situation is particularly serious with an estimated two million infected persons (Fig. 22.4). In Central Africa, HIV-1 was probably spreading during the late 1970s and early 1980s, at about the same time as in the USA. In small retrospective serological surveys in pregnant women in Nairobi there were none with anti-HIV antibody in 1980, whereas 2 per cent of such women were seropositive in 1985. As many women as men are infected, indicating that heterosexual spread is a major

Table 22.4 Adult AIDS cases in the USA: percentage of cases in various exposure categories

Exposure category	Male (Total 100 000 cases)	Female (Total 10 000 cases)
Male homo/bisexual	67	
Intravenous drug use (heterosexual)	18	52
Male homo/bisexual contact and i.v. drug use	8	
Haemophilia/coagulation disorder	1	0
Heterosexual contact	2	31
Recipient of blood transfusion	2	10
Undetermined	2	7
Total	100	100

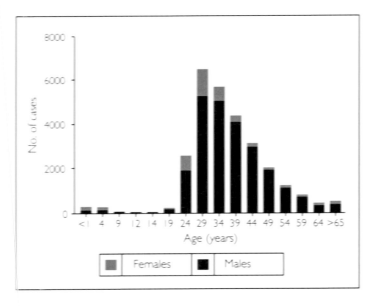

Fig. 22.3 The relationship between age and incidence of HIV-1 infection. Except in babies infected *in utero* and haemophiliacs, the highest incidence of infection coincides with the ages of maximum sexual activity.

factor. But contributory factors such as the use of unsterilized needles and concomitant immunosuppressive infections with parasites such as malaria and widespread genital ulcers — which offer a route of entry of the virus — may be significant in Africa.

In West Africa another virus, HIV-2 (Fig. 22.5), has been isolated from persons at special risk for sexually transmitted disease, such as prostitutes; but it has not been associated with such severe clinical illness as HIV-1.

Table 22.5 Individual action for the prevention of AIDS

Needle-borne transmission

Needles or syringes must not be shared

Sexual transmission

A reduction in the number of sexual partners to decrease the chance of being exposed to an infectious person. The following guidelines can be used to decrease the risk of infection

Absolutely safe

Mutually monogamous relationship

Very safe

Non-insertive sexual relations using a condom containing spermicide

Risky

Anything else

Perinatal transmission

Persons exposed to HIV should have the antibody test. Antibody-positive women should not become pregnant

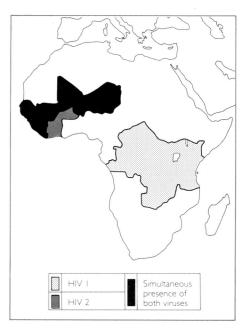

▒	HIV 1	■	Simultaneous presence of both viruses
■	HIV 2		

Fig. 22.4 Geographical distribution of HIV-1 and HIV-2 in Africa. The prevalence of HIV-1 is highest in the eastern areas, whereas HIV-2 is localized in the west.

How long has AIDS been with us?

As we have mentioned, AIDS first came to notice in 1981 but, when serological techniques for diagnosing it became available, retrospective tests on stored blood samples revealed that HIV infections were occurring unnoticed during the 1960s. Recently, workers at the University of Manchester turned the clock back still further. They had previously reported the death, in 1959, of a sailor from cytomegalovirus and pneumocystis infections. This combination was a puzzle at the time, but is now, of course, known to be characteristic of AIDS. Accordingly, they examined stored tissues by the highly sensitive polymerase chain reaction (Chapter 27) and detected HIV proviral DNA. In view of the long incubation period, this finding places the earliest verifiable infection in the UK at some time in the 1950s.

Public Health regulations and HIV-1

National health authorities throughout the world are reacting rather differently to the threat of an epidemic of HIV-1. Certain countries, including the USA and the CIS (ex-USSR), are, against the advice of the World Health Organization, banning long-stay visitors who have antibody to HIV-1. In some European countries, such as Iceland, Sweden, and the UK, the health authority can restrict the movement of a case of AIDS in a 'dangerously infectious state', but AIDS has not joined the group of notifiable diseases in England and Wales. An application may be made by the local authority to a Justice of the Peace to examine forcibly and detain an AIDS patient in hospital, but these provisions are intended for use only in exceptional circumstances.

HTLV-I

In the subtropical south-western part of Japan, particularly the islands around Okinawa, the prevalence of HTLV-I antibody is as high as 15 per cent; the rates increase with age and are higher in men than in women. Studies on people who have emigrated from this area suggest that the incubation period of clinical illness is 15–20 years. In Japan itself, the risk of an antibody-positive person developing ATLL is about 1/1000. In Jamaica, the prevalence of antibody is about 6 per cent and, in expatriate Caribbean communities, 1–4 per cent; the rates are similar in West African countries.

3.6 Laboratory diagnosis of AIDS

Serological tests

Many commercial kits are available for the serological diagnosis of HIV-1 infection. Some serological methods can also detect HIV-2 and HTLV-I infections. The most commonly used kits detect antibodies to HIV-1 env protein by an ELISA method (see Chapter 27). In view of the great personal and social implications of a positive test for anti-HIV antibody, it is important to do a confirmatory test by a completely different technique such as immunofluorescence, Western blotting, or by an ELISA test for antibody to another viral gene product, namely gag protein. There may be a period of several months when the virus is present but the infected person has not yet produced detectable antibodies. If, therefore, a person is a member of one of the high-risk groups, serological testing will have to be performed again after 6 or even 12 months before a negative result is acceptable.

Sensitive virus antigen capture tests are widely used to detect free viral p24 antigen in the blood of infected persons (see Chapter 27). Figure 22.2 illustrates the sequential development of antibodies and p24 antigenaemia and these tests have some prognostic value. The quantification of p24 antigenaemia is useful to monitor the efficacy of antiviral drugs; for example, following administration of zidovudine the virus p24 antigen level in the serum drops sharply.

An important laboratory diagnostic test is establishment of the ratio of CD4 to CD8 lymphocytes — normally about 2:1 — which is greatly depressed in persons with clinical manifestations of AIDS. Finally, microbiological tests for opportunistic infections, e.g. bronchoscopic washings for *Pneumocystis carinii*, play an important role in the diagnosis of AIDS and the continued monitoring of the patient's progress.

Virus isolation

Only specialized laboratories undertake the still difficult task of isolating virus from the preferred target cell, the human lymphocyte. A category III laboratory, with restricted access and appropriate safety cabinets, is the minimum safety requirement for viral isolation. Lymphocytes from persons under investigation are co-cultivated either with artificially stimulated lymphocytes from uninfected individuals or with lymphocyte cell lines and assayed subsequently for RT, virus-induced giant cell formation, or, by ELISA, for p24 gag antigen. Several weeks are needed for these assays. Virus can be isolated from up to 80 per cent of clinically ill persons, but at the terminal stages of the disease its recovery becomes more difficult. HIV may also be isolated from saliva, tears, urine, breast milk, and semen, but the frequency of recovery is much lower than from lymphocytes, presumably because these fluids contain less infective virus.

Safety aspects of sample collection

The clinical sample. Gloves and a disposable plastic apron or gown must be worn when taking blood or other specimens from ARC or AIDS patients. Eye protection is recommended, particularly in view of the case of a nurse infected after splashing blood in her eye. Particular care must be taken not to contaminate the outside of the specimen container. Disposable needles and syringes must be used for blood collection.

Transport. Samples for virus isolation or serology must be transported to the laboratory in a properly labelled plastic bag in robust screw-capped containers with a hazard warning label. Pins, staples, or metal clips must not be used to seal the bag because of the danger of skin pricks.

Accidents must be reported and skin pricks treated immediately by encouraging bleeding and washing with soap and water. Some hospital pharmacies retain a stock of the antiviral drug zidovudine for immediate administration to staff who have been potentially infected via skin pricks.

4 REMINDERS

◆ Retroviruses are enveloped and contain **diploid RNA**; they possess a **reverse transcriptase (RT)** enzyme able to catalyse transcription of viral RNA into DNA (proviral DNA).

◆ The lentiviruses such as HIV-1 and HIV-2 have unique controlling genes, *tat*, *rev*, *vif*, and *nef*.

◆ **HTLV-I** causes cancer (**adult T-cell leukaemia lymphoma, ATLL**) and also a chronic neurological disease called **tropical spastic paraparesis (TSP)** or HTLV-I-associated myelopathy (HAM). **HTLV-II** causes a rare **hairy cell leukaemia**.

◆ Members of the lentivirus group such as **HIV-1** and **HIV-2** cause **slow persistent infection** with viral gene integration into the DNA of CD4-positive lymphocytes. A substantial proportion of those infected develop the **acquired immune deficiency syndrome (AIDS)**.

◆ The three stages of HIV-1 disease are: (1) the **asymptomatic phase**; (2) **AIDS-related complex (ARC)** with persistent lymphadenopathy, night sweats, and diarrhoea; and (3) full-blown **AIDS** with a plethora of opportunistic infections. Paediatric AIDS may be more difficult to diagnose clinically, especially in developed countries, because of concomitant problems of drug addiction.

◆ Laboratory studies are required to confirm infection with HIV-1, HIV-2, or HTLV-I. ELISA detects **specific antibodies** to these retroviruses. Determination of amounts of **circulating viral p24 antigen** is an important prognostic test of progress of the disease. Virus isolation is not performed routinely.

◆ The life of AIDS patients may be prolonged by treatment with the antiretroviral drug **zidovudine** and prompt treatment of bacterial, parasitic, and viral opportunist infections. No vaccines are available.

◆ Special risk groups for HIV-1 and HIV-2 infection in the community are **male homosexuals**, **intravenous drug abusers**, **sexual partners of the above two groups**, and **babies of infected mothers**.

◆ The virus is now spreading among the heterosexual community.

◆ The individual can take important actions to prevent infection both by using safe sex procedures and by not sharing needles and syringes.

CHAPTER TWENTY-THREE

UNCONVENTIONAL AGENTS: VIROIDS, VIRINOS, AND PRIONS

1 INTRODUCTION

The title of this chapter may convey the impression that we are venturing into subatomic physics; and perhaps the people who thought up names for these strange infective agents were influenced by their very small size.

Viroids are small, circular, single-stranded RNA molecules, that do not apparently code for polypeptides. They cause disease in a variety of plants, but not, so far as is known, in animals. They need not therefore concern us further, except inasmuch as, like the other agents to be discussed here, their mode of replication remains a mystery. **Virino** is the term applied to a hypothetical particle made up of a small, non-coding nucleic acid molecule associated with a protein of host origin. The notion of **prions** — proteinaceous infectious particles — is even more puzzling, since such agents possess no nucleic acid at all. But however unlikely these concepts appear, they have been developed in response to very real clinical and laboratory findings that cannot be explained in terms of conventional viruses. Although the evidence for prions has recently become much firmer, some workers are still reluctant to use this or other specific terms and employ the non-committal designation 'unconventional agents' as causes of the diseases now to be described. We shall refer to them here as **scrapie-like agents (SLA)**, after the disease of sheep that is the best-studied of this group.

2 DISEASES CAUSED BY SCRAPIE-LIKE AGENTS

In 1920 and 1921 respectively, Creutzfeldt and Jakob described the first cases of the syndrome that now bears their names. Forty years later a somewhat similar illness known as kuru was described in the Fore tribe of Papua New Guinea; both diseases were subsequently linked with scrapie, an infection of sheep, and transmissible encephalopathies affecting mink and cattle. These unlikely partners have in common **very long incubation periods, a progressive and fatal disease confined to the CNS**, and **spongiform (vacuolated) degeneration** of the neurons accompanied by a **reactive astrocytosis**.

2.1 *Infections of humans* (Table 23.1)

Creutzfeldt–Jakob disease (CJD) has an incubation period of several years, sometimes as many as 20 or 30. It is characterized by dementia and **recurrent seizures**; death usually occurs within a few months of onset. It is a rare condition, only about 20 cases a year being reported in the UK, mostly in middle-aged or elderly people. The incidence — of the order of one case per million people *per annum* — is similar in other countries, the exception being Israeli Jews originating from Libya, in whom it is substantially higher. There is also a degree of familial clustering, suggesting that **genetic factors** may sometimes influence susceptibility to the infection. This is certainly true of a similar but even more rare disease, the Gerstmann–Straussler syndrome.

In most cases of CJD the mode of transmission is unknown. There have, however, been a number of **iatrogenic infections**, transmitted unwittingly from people who were incubating the disease. These infections resulted from transplants of corneas or dura mater, inadequately sterilized neurosurgical instruments, or contamination of batches of human growth hormone.

Table 23.1 Diseases caused by unconventional agents: spongiform encephalopathies of humans

Disease	Mode of transmission
Creutzfeldt–Jakob disease	
Sporadic cases	Unknown
Iatrogenic cases	Inadequately sterilized neurosurgical instruments; certain transplants from donors incubating the infection; infected pituitary growth hormone
Kuru	Ritual cannibalism, via contamination of cuts and abrasions on skin with infected brain tissue

Measures are now taken to guard against these known hazards, but it is possible — although perhaps unlikely — that there are others of which we are as yet unaware.

Kuru is also a progressive and fatal dementia, usually with pronounced cerebellar involvement leading to difficulty in walking. This curious disease was limited to the Fore tribe; it was spread by ritualistic cannibalism, during which the brains of relatives were eaten by the women and children only, among whom there was a very high incidence. It is thought that the infection was transmitted by contact of tissue with cuts and abrasions, rather than by ingestion. When cannibalism was stopped in the 1950s, kuru started to disappear, although, because of the long incubation period, there is still an occasional case. It seems that this disease is of comparatively recent origin; there is speculation that it may have been introduced as a result of butchering someone with CJD, or who was incubating it.

2.2 *Infections of animals* (Table 23.2)

Scrapie, an enzootic infection of **sheep and goats**, is the best studied of all the spongiform encephalopathies, and from it we have gained most of our knowledge of the unconventional agents. The name derives from the need of infected animals to rub themselves against objects, e.g. fences, to try and relieve the characteristic intense itching. Infection is transmitted directly or indirectly from animal to animal and is often acquired at birth. The incubation period is 1–5 years; the onset of itching and ataxia heralds a downward, invariably fatal course lasting a few months. There is good evidence of genetic variation in susceptibility to the infection and also for the existence of different strains of scrapie agent. Scrapie is prevalent in the UK, particularly in the north of England and Scotland, but some countries, e.g. Australia, are completely free of this infection.

Table 23.2 Diseases caused by unconventional agents: spongiform encephalopathies of animals

Disease	Mode of transmission
Scrapie	
Adult animals	Contact or grazing on infected pastures
Newborn animals	Perinatal transmission from ewe
Transmissible mink encephalopathy (in captive animals)	In foodstuffs (?)
Chronic wasting disease of deer (in captive animals)	Unknown
Bovine spongiform encephalopathy	Foodstuff (meat and bonemeal concentrates)

Transmissible mink encephalopathy (TME) and chronic wasting disease of deer are both scrapie-like diseases occurring in captive animals.

Bovine spongiform encephalopathy (BSE) is the latest of these syndromes to be recognized and is giving some cause for concern in view of its potential — slight though it may be — for being transmitted to humans. It was first recognized in the mid-1980s and so far is most widespread in the UK, where some thousands of cattle have been affected. BSE has also been reported in mainland Europe and the Channel Islands. It manifests as disturbances of behaviour — the 'mad cow disease' of the popular press — and of motor function. The pathological changes in the brain are very similar to those of scrapie. To all intents and purposes, this disease is scrapie of cattle.

The most likely explanation of its sudden appearance is that a scrapie agent — possibly a mutated strain — was transmitted to cattle by way of infected food concentrates made from meat and bonemeal. There is evidence that outbreaks are related to the type of extraction process used for preparing these cattle feeds.

3 EXPERIMENTAL TRANSMISSION TO OTHER SPECIES

Kuru has been transmitted to a number of species of primates and CJD both to primates and other animals. Perhaps the most important finding was that mice and Syrian golden hamsters can be artificially infected with scrapie; the incubation period is only a matter of months, thus making practicable infectivity titrations and other studies, including research on pathogenesis.

4 PATHOGENESIS AND PATHOLOGY

Studies of the pathogenesis of scrapie in mice are important because of the light they may shed on spongiform encephalopathies in humans. After a brief viraemia, the agent first multiplies in the spleen and other organs of the reticuloendothelial system, from which it spreads to the CNS by migrating up peripheral nerves to the spinal cord.

The histopathology of scrapie, CJD, and kuru is very similar. In addition to the spongiform (vacuolated) degeneration (Fig. 23.1) there is astrocytosis but little or no inflammatory reaction. Amyloid plaques are sometimes formed

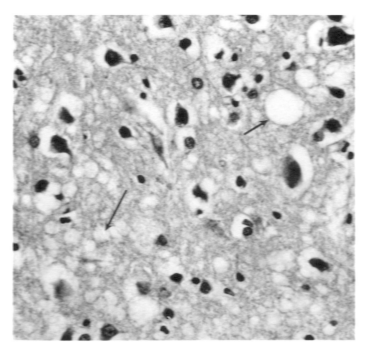

Fig. 23.1 Section of brain from case of Creutzfeldt–Jakob disease. The 'holes' (arrowed) are characteristic of all the spongiform encephalopathies. (Courtesy of Dr Carl Scholtz, The London Hospital Medical College.)

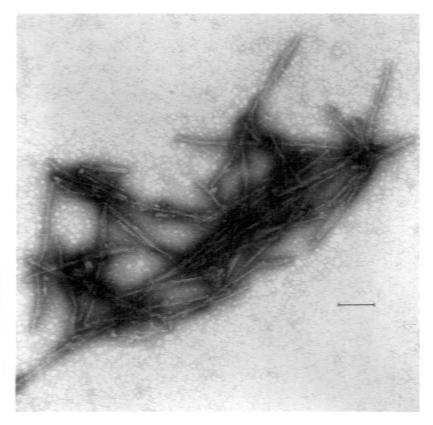

Fig. 23.2 Electron micrograph of scrapie-associated fibrils (SAF). From brain of scrapie-infected mouse. (Courtesy of Dr J. Hope, AFRC and MRC Neuropathogenesis Unit.) Scale bar, 100 nm.

in neural tissue. In scrapie, BSE, and CJD the most characteristic feature is the presence of **scrapie-associated fibrils (SAF)**. These are detectable by electron microscopy of brain extracts (Fig. 23.2) and appear to be formed from a protein, PrP, that is a normal component of cells but which in SLA infections is modified in its physicochemical characteristics.

5 SPECIFIC THERAPY

The difficulties in mounting controlled drug trials in such a rare disease as CJD are obviously formidable, and, although some workers have reported that certain substances, e.g. amphotericin B and dextran sulphate, partially inhibit scrapie in mice or hamsters, there is as yet no confirmed report of an antiviral molecule that is useful against CJD in humans.

6 THE NATURE OF THE SCRAPIE-LIKE AGENTS

Any explanation of the nature of SLA must take account of the following facts.
1. They appear to contain little or no nucleic acid, and certainly not in quantities sufficient to code for virus-like structure and functions.
2. Nevertheless, SLA are infective, i.e. can reproduce themselves.
3. Some at least can undergo genetic variation in properties.
4. The ability of some SLA to cause disease is related to genetic factors in the host.

The reconciling of these apparently discordant features is still beset by controversy. The view that seems to be emerging as the front runner, well backed by experimental evidence, is that of Stanley Prusiner in the USA, who devised the term 'prion'. In brief, prions are an abnormal form of the protein PrP, which itself is encoded by a gene on human chromosome 20 (mouse chromosome 2); SAF are composed of this modified form, known as PrP^{Sc}. One suggestion among many is that PrP^{Sc}, by alternately combining with and parting from PrP molecules, induces conformational changes in the latter that convert them to the modified form at an exponential rate. Another is that a small host-specified nucleic acid might act as a cofactor in the conversion process.

7 SAFETY MEASURES

Since there are no laboratory diagnostic tests and no treatment for these infections, we must for the time being concentrate on means for preventing their spread, whether by natural means or as the result of accidents.

7.1 *Creutzfeldt–Jakob disease*

We have seen that unconventional agents are **highly resistant to the usual disinfection procedures** and that there have been a number of instances of transmission from one patient to another by inadequately sterilized neuro-surgical instruments. It must be emphasized, however, that there is no documented instance of health care staff becoming infected by day-to-day contact with CJD patients, or as a result of surgical procedures not involving the CNS. But in the light of past iatrogenic transmissions, special care must be taken when undertaking operations or autopsies on patients likely to have CJD. In the UK, the Department of Health and Social Security has issued guidelines for safety, including recommendations for **extended autoclaving of instruments used in neurosurgical procedures** (See Appendix A). Immersion in 1 N sodium hydroxide has also been recommended, but, with the possible exception of this and strong hypochlorite, none of the conventional chemical disinfectants can be relied upon. It should be noted that the CJD agent can survive for years in brains preserved in formalin.

7.2 *Scrapie and bovine spongiform encephalopathy*

If scrapie can be transmitted to cattle, why not to man? All we can say at present is that there is no known association between the presence of sheep scrapie in a particular area and the incidence of CJD, which is similar in countries with and without scrapie. Furthermore, CJD has been reported in people who have never eaten sheep meat. Nevertheless, the authorities in the UK at present regard the possibility of transmission of BSE from cattle seriously enough to make it a notifiable disease and to order the slaughter of any animals showing signs of it. Furthermore, the consumption of infected carcasses is banned, as is the use of bovine meat and bonemeal in animal foodstuffs. Nevertheless, this policy dodges the issue of whether — as seems likely — animals are infective during the incubation period. It was hoped that BSE would prove not to spread from animal to animal and would disappear of its own accord if infection of their foodstuffs could be prevented. However, a recent report that BSE has apparently been transmitted from a cow to her calf casts a cloud on this optimistic view. Clearly, however, a close watch will have to be kept for possible cause-and-effect relationships between animal spongiform encephalopathies and similar syndromes in humans. The suggestion has been made that SLA may also be implicated in Alzheimer's disease, a progressive form of dementia seen in elderly people, which is also characterized by the presence of tangles of fibrils in the brain.

8 REMINDERS

◆ The spongiform encephalopathies have in common:
   1. very long incubation periods;
   2. a progressive, uniformly fatal course;
   3. characteristic spongiform changes in the brain;
   4. lack of inflammatory and immune responses.

◆ They occur in humans, (**kuru**, **Creutzfeldt–Jakob disease**, or CJD) and a variety of animal species, notably sheep (**scrapie**) and ruminants (**bovine spongiform encephalopathy**, or BSE).

◆ These syndromes are caused by infective agents, sometimes called unconventional viruses, of which the best studied is that of scrapie. Apart from the finding of characteristic fibrils (**scrapie-associated fibrils**, or **SAF**) in brain extracts, there is little by which to identify these scrapie-like agents (SLA).

◆ The nature of SLA is not as yet understood. Their remarkable **resistance to chemical and physical agents**, including irradiation, argues against the presence of a nucleic acid genome, but the existence of different strains suggests that they make use of some genetic mechanism.

◆ Scrapie can be transmitted experimentally to mice, in which most of the studies of infectivity and pathogenesis have so far been done.

◆ **Bovine spongiform encephalopathy (BSE)** affects cattle, so far only in the UK. The probability that it arose from sheep scrapie raises the possibility that it could be transmitted to humans by ingestion of infected meat or offal, but so far there is no evidence that this occurs.

◆ In the absence of specific means of diagnosis or treatment, efforts are directed toward preventing accidental infection of health care workers with CJD, and transmission of BSE by way of contaminated foodstuffs of animal origin.

SPECIAL
SYNDROMES

CHAPTER TWENTY-FOUR

VIRAL DISEASES OF THE CENTRAL NERVOUS SYSTEM

1 INTRODUCTION

If you are reading these chapters in sequence, you will by now have realized that many of the viruses so far dealt with can cause disease of the central nervous system (CNS). In fact, members of nearly all the 15 or so families of animal viruses can do so in one way or another. This propensity varies however from family to family: in some, for example the *Herpesviridae*, all members are potentially neurotropic; by contrast, few of the *Papovaviridae* cause CNS disease and then only rarely. Manifestations of CNS involvement caused by individual viruses are described in the appropriate chapters; here, we shall deal with the topic in a more general way so as to give you an overall picture of the kinds of lesions encountered and of the principles of laboratory diagnosis.

At first sight, the number of viruses involved and the variety of syndromes are confusing. We can simplify things by classifying these illnesses into three groups (Table 24.1). Group 1 is straightforward, comprising acute infections that directly involve the CNS; group 2 illnesses are also acute, but are less directly related to the associated virus infections; and the chronic fatal diseases in group 3, fortunately all rare, are caused by grossly abnormal responses to conventional viruses, or by virus-like agents whose nature is not yet understood.

A general point to remember is that the ways in which the CNS can respond to virus infections are limited: the neurons themselves may undergo **lysis**, **demyelination**, or **spongiform degeneration** and the inflammatory response is mainly expressed by the appearance of **lymphocytes** and activation of **microglia**, the macrophages of the CNS. These inflammatory cells tend to

Table 24.1 Virus diseases of the central nervous system (CNS): general classification

	Virus usually demonstrable in CNS	
	Yes	No
Acute	Group 1	Group 2
Chronic	Group 3	

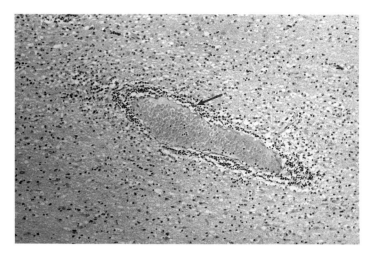

Fig. 24.1 Section of brain from a case of herpes simplex encephalitis. Note the 'cuffing' of the blood vessel with mononuclear cells (arrowed) and the intense proliferation of microglia.

collect in the perivascular spaces of the small blood vessels, an appearance graphically described as 'cuffing' (Fig. 24.1). Another important response of great diagnostic value is the manufacture of **specific IgM antibody** within the CNS; since such antibody cannot cross the 'blood–brain barrier', its presence in the cerebrospinal fluid is good evidence of an active infection by the corresponding virus **within** the CNS compartment.

A final introductory remark: please note the word 'predominant' in the headings of Tables 24.2–24.4. Like most clinicopathological syndromes, these are often not clear-cut and allowance must be made for variations on the main themes. For example, in those illnesses classified as 'meningitis' there may also be some involvement of the brain substance.

2 ACUTE INFECTIONS
(GROUP 1, TABLE 24.2)

These syndromes all arise from direct invasion of the CNS by the virus concerned, which can usually be isolated from the CSF or brain.

2.1 *Viral meningitis*

During some virus infections of childhood, notably with **mumps** and some **enteroviruses**, there may be a degree of meningitis indicated by headache,

Table 24.2 Virus diseases of the central nervous system. Group 1: acute infections

Predominant syndrome	Viruses	Predominant neurological lesions
Meningitis	**Enteroviruses**, especially ECHO, Coxsackie A and B, enteroviruses 70 and 71, poliovirus	Inflammation of the meninges, with or without some degree of encephalitis
	Mumps, lymphocytic choriomeningitis, louping ill, Epstein–Barr virus, HSV-2, VZV	
Poliomyelitis	Polioviruses; occasionally other enteroviruses	Meningitis; lysis of lower motor neurons
Meningoencephalitis	HSV-1, arboviruses	Necrosis of neurons in grey matter of brain
Encephalitis	Rabies	Varying degrees of neuronal necrosis; perivascular and focal inflammation
AIDS dementia complex (ADC)	HIV-1	Meningitis; cortical atrophy; focal necrosis, vacuolation, reactive astrocytosis, and microgliosis in subcortical areas, peripheral neuropathy

neck rigidity, fever, and sometimes vomiting. In adults, meningitis is also an occasional feature of genital tract infection with herpes simplex virus type 2. These episodes, although somewhat alarming, usually resolve quickly and have no sequelae. They are often referred to as 'lymphocytic' or 'aseptic' meningitis because of the predominance of lymphocytes in the cerebrospinal fluid (CSF) although some polymorphs may be present in the early stages. On lumbar puncture, the opening pressure is normal or slightly raised and the fluid is clear, as opposed to the turbid CSF of pyogenic meningitis. There are several hundred cells per ml, the concentration of protein is normal or slightly raised (not above 100 mg/100 ml) and the sugar > 40 mg/100 ml. The associated virus can usually be isolated from the CSF. It is worth remembering that lymphocytic choriomeningitis, an arenavirus, produces a similar clinicopathological picture; it is acquired from animals, usually mice or hamsters, and should come to mind in case of an outbreak of meningitis in laboratory workers handling animals. Louping ill, another zoonosis, is the only arbovirus found in the UK and occasionally causes meningitis in those in contact with sheep, particularly in northern England and Scotland.

In any case of lymphocytic meningitis the possibility of **tuberculosis** must be thought of, especially if the sugar concentration in the CSF is depressed; failure to diagnose such an infection could be catastrophic. **Leptospirosis** is another non-viral and treatable cause of lymphocytic meningitis.

2.2   *Poliomyelitis*

This syndrome is fully described in Chapter 14. It is classified separately because of the selective effects on the anterior horn cells.

2.3   *Meningoencephalitis*

Herpes simplex virus (HSV) infections

By contrast with simple meningitis, usually associated with HSV-2, the much more dangerous meningoencephalitis is nearly always caused by HSV-1 (see Chapter 15). Early diagnosis is important, because treatment at the earliest possible moment with **acyclovir** considerably reduces both mortality and the incidence of sequelae. Indeed, this drug is often started as soon as a reasonable suspicion of HSV encephalitis arises, without waiting for any test results, but its availability and lack of toxicity must not be allowed to dull alertness to the possibility of a quite different diagnosis.

The pathological lesions are those of an **acute necrotizing encephalitis**, characteristically most pronounced in the **temporal lobes** and accompanied by intense meningeal inflammation. There is usually severe cerebral oedema which may mimic the clinical presentation of a space-occupying lesion. This is the classical picture, but less severe cases probably occur that are never correctly diagnosed.

The laboratory diagnosis depends on

◆   detection of specific **antibody** in the CSF (see Section 5);
◆   detection of **virus** in a brain biopsy by culture, presence of typical inclusion bodies, immunofluorescence, or electron microscopy.

Of these methods, the second is the most rapid and definitive and has the additional advantage that it may establish an alternative diagnosis, but its use is somewhat controversial and is, of course, not possible in the absence of adequate surgical and laboratory facilities.

Recently, Swedish workers described a test which may resolve the uncertainties of diagnosing HSV infection of the CNS. A polymerase chain reaction test (PCR; see Chapter 27) proved highly sensitive and specific for detecting HSV DNA in cerebrospinal fluid; such a method may eventually come into routine use, but great care will have to be taken to avoid contamination of samples with minute traces of HSV DNA from persons handling them.

Treatment with acyclovir must be adequately backed up by intensive care, including energetic measures to reduce intracranial pressure.

Arboviruses

These tropical infections are described in Chapter 17. Meningoencephalitis is a major feature in many of them (Table 17.3) but its severity varies considerably and the incidence of sequelae differs according to the virus involved. The clinical features in the acute stage are similar, comprising fever, headache, and meningismus; in severe cases there may be convulsions proceeding to coma and death. Laboratory identification depends on tests for serum antibody, but the diagnosis is usually made on clinical and epidemiological criteria. There is no specific treatment for any of these infections, although vaccines are available for preventing some of them, e.g. Japanese B encephalitis.

2.4 *Encephalitis*

The only example of a 'pure' encephalitis without meningeal involvement is provided by **rabies**, which is described in detail in Chapter 16. By contrast with the other infections in this group, which spread to the CNS via the bloodstream, rabies gains access to the cord and brain by migrating up the peripheral nerves serving the site of the injury.

2.5 *AIDS dementia complex (ADC)*

The neurotropism of HIV was at first overshadowed by the associated intercurrent infections with micro-organisms that may also affect the CNS such as herpesviruses, mycobacteria, and toxoplasma. It is now realized, however, that this virus is the direct cause of brain damage resulting in a variety of symptoms and signs, including particularly psychiatric problems and motor and sensory disturbances (see Chapter 22). These have to be disentangled from the effects of opportunistic infections, so that differential diagnosis depends on the results of careful clinical examination supported by a battery of laboratory tests including computerized tomography of the brain. In ADC, the CSF protein is usually elevated and lymphocytes may be present; virus can be isolated from the fluid. The histological appearances are summarized in Table 24.2. Viral antigen is detectable in the white matter of the brain and basal ganglia and is associated particularly with multinucleate cells and macrophages.

3 ACUTE POSTEXPOSURE SYNDROMES
(GROUP 2, TABLE 24.3)

We now come to a group of three syndromes that have three points in common.

◆ They are each associated with exposure to any one of several viruses.
◆ The associated viruses cannot be isolated from the CNS.
◆ The pathogenesis is obscure.

Table 24.3 Virus diseases of the central nervous system. Group 2: acute postexposure syndromes

Syndrome	Viruses	Predominant neurological lesions
Postinfection and postimmunization encephalomyelitis	Measles, rubella, mumps, varicella-zoster, rabies (killed neural tissue vaccines), vaccinia	Demyelination; pronounced microglial proliferation and lymphocytic 'cuffing'
Guillain–Barré syndrome	Many viruses, including influenza (vaccine), enteroviruses, Epstein–Barr virus, cytomegalovirus	Polyradiculoneuritis with demyelination, inflammation, and degeneration of nerve roots and ganglia
Reye's syndrome	Influenza B, varicella, adenoviruses	Non-inflammatory encephalopathy; cerebral oedema

We have coined the term 'postexposure syndromes' for this group, because they follow exposure to both live and inactivated viruses or viral antigens; the usual expression, 'postinfection' is thus inadequate for the group as a whole.

3.1 *Postinfection and postimmunization encephalomyelitis*

In 1905 reports started to appear of mysterious paralytic illnesses following immunization with both rabies and smallpox vaccines. The incubation periods were similar and remarkably constant: 14–15 days after the first dose of rabies vaccine and 10–13 days after smallpox vaccination. Later, it was noticed that there were occasional similar episodes about 2 weeks after onset of the common virus infections of childhood.

It took some time to distinguish these syndromes from other untoward reactions to the vaccines and from the meningitis sometimes associated with these childhood infections, both of which tended to occur earlier after immunization or the onset of illness. Eventually, however, a clearer picture emerged. The illness following rabies immunization presents either as an ascending Landry-type paralysis with a mortality of about 30 per cent, a more frequent dorsolumbar myelitis with a mortality of around 5 per cent, or a comparatively benign neuritis affecting the cranial nerves. The syndrome associated with smallpox vaccination and the childhood fevers was rather that of meningoencephalitis but in all these conditions the histological appearances are those of **demyelination, microglial proliferation**, and **cuffing** of the blood vessels. It must, however, be pointed out that, although the diagnosis of postimmunization encephalomyelitis is fairly straightforward, the clinical distinction between meningitis complicating the childhood infections and true postinfection ence-phalomylitis is often blurred.

How frequent are these episodes? This question raises the interesting problem of geographical association. The incidence of postvaccinial encephal-itis was notoriously higher in Holland than elsewhere, about 1/4000 vaccina-tions compared with, for example, 1/50 000 in England and Wales. This difference was not attributable to the strain of vaccinia in use and seemed to depend on some environmental factor that remains totally obscure.

Because it is cheap, rabies vaccine prepared from neural tissue is still widely used in Third World countries. The reported incidences of neuropara-lytic episodes are of the same order as those for smallpox vaccine, varying from 1/1000 to 1/10 000 courses of vaccine.

Postinfection encephalomyelitis is most common after **measles**, occur-ring in perhaps 1/1000 cases; it is less frequent after chickenpox, rubella, and mumps.

Clinical aspects

The onset is often abrupt and is heralded by convulsions or coma. Personality and behavioural changes are frequent and may be followed by any of a wide range of focal neurological signs, mostly motor. The severity varies consider-ably; the prognosis is poor when coma supervenes and neurological defects following recovery are often severe.

The pathogenesis of these conditions remains a mystery. The neuropar-alytic accidents following rabies vaccination occurred with both live attenuated and killed vaccines prepared from the brains of infected animals, but are not associated with modern vaccines made in cell culture, which might suggest that injection of neural tissue is the predisposing factor; indeed, a similar disease can be induced by injecting rabbits with rabbit-brain extract. But this does not explain the postvaccinial and postinfectious syndromes. Since failure to detect the relevant viruses in the CNS seems to rule out any direct effect of viral multiplication, the activation of some unknown latent virus has been suggested, or, more plausibly, that these episodes are the result of an abnormal immune response to the viral antigen.

Apart from identifying the causal infection, laboratory tests are not particularly helpful. The CSF shows the now familiar picture of raised protein, normal sugar, and lymphocytosis. As we have seen, virus cannot be isolated from the CNS of these cases.

3.2 Guillain–Barré syndrome

This is another demyelinating disease that has features in common with the syndromes described in Section 3.1. It differs in affecting the **nerve roots** rather than the substance of the brain and cord and follows exposure to a number of viruses (Table 24.3), which again need not be live. This was dramatically illustrated by cases of Guillain–Barré syndrome in the USA that followed mass vaccination in 1976 against a newly appeared H1N1 strain of influenza virus (Chapter 10). The incidence of this syndrome was 1 per 130 000 of those vaccinated, four to eight times higher than in the general population. The reason why this inactivated vaccine—and not others prepared subsequently—should have had such dire effects remains a mystery.

In Guillain–Barré syndrome there is no pleocytosis in the CSF, but the protein concentration is well above normal.

3.3   *Reye's syndrome*

This differs from the postexposure syndrome so far described, since it presents as a **non-inflammatory encephalopathy** with **cerebral oedema** and **fatty degeneration of the liver**. It affects children, in whom the mortality rate is 25–50 per cent. Survivors may have neurological damage. Again, the pathogenesis is unknown, but there has been much discussion, so far unresolved, about the possible role of virus infections—notably influenza B, varicella, and adenoviruses—as precipitating factors. It has been suggested that salicylates may also be implicated, especially if given during the course of an influenza B infection. The current advice, which seems sensible, is not to give aspirin to children with febrile illnesses.

4   CHRONIC INFECTIONS
(GROUP 3, TABLE 24.4)

As with the illnesses in group 1, the causal agents can be demonstrated in the CNS, but there the resemblance ends, because these syndromes are all very rare and uniformly fatal. Their pathogenesis presents as many puzzles as those just discussed in group 2.

4.1   *Subacute sclerosing panencephalitis (SSPE)*

Because of the comparatively young age group affected, its duration, and relentless downhill course, this must surely rank as one of the most distressing virus infections of the CNS. Around 7 years after contracting measles (or, more rarely, congenital rubella) a child starts to develop neurological signs. At first, there may be nothing more than a slight impairment of movement in one limb or another, or behaviour disorder. Within weeks or months,

Table 24.4 Virus diseases of the central nervous system. Group 3: chronic infections (all rare)

Syndrome	Viruses	Predominant neurological lesions
Subacute sclerosing panencephalitis (SSPE)	Measles, rubella	Neuronal degeneration; demyelination; microglial proliferation
Progressive multifocal leucoencephalopathy (PML)	Papovaviruses (JC, very rarely SV40)	Multiple foci of demyelination in brain; hyperplasia of oligodendroglia and astrocytes
Creutzfeldt–Jakob disease (CJD); kuru	Scrapie-like infective agents ('unconventional' viruses)	Spongiform degeneration and atrophy of brain and anterior horn cells; astrocytosis

however, the true nature of the condition becomes apparent, with mental deterioration, myoclonic seizures, spastic paralyses, and blindness. Death occurs within 1 to 3 years from the onset.

The incidence varies with country and with ethnic group. As a very rough guide, there is about one case per million of acute measles, but this figure is rather higher in Scandinavia and lower in the UK and USA. Children who acquire measles below the age of 2 years are at greatest risk, and males are more often affected than females.

The pathology is curious. There is generalized encephalitis with the usual cuffing of blood vessels with mononuclear cells. Glial proliferation is pronounced and both glia and neurons contain **inclusion bodies**. These consist of randomly arranged measles virus nucleocapsids, which do not infect cell cultures in the normal way, but can be made to express replicating virus by co-cultivating biopsied brain cells with other cells susceptible to measles infection. There is a **defect in the production of the measles virus M protein** (Chapter 9) which accounts for the difficulty in replication.

The diagnosis is readily confirmed by the finding in the serum of a **high titre of antimeasles antibody, including IgM**, and in the CSF of **measles IgG antibody lacking the normal anti-M component**.

4.2 *Progressive multifocal leucoencephalopathy (PML)*

Patients with lymphoproliferative disorders and other immunosuppressive conditions occasionally develop a progressive and fatal illness with gross intellectual impairment and paralyses. This syndrome is caused by a polyomavirus and is described in Chapter 21.

4.3 *Subacute spongiform encephalopathies*

These are rare, progressive, uniformly fatal infections caused by virus-like agents that have not yet been fully characterized. They are dealt with in Chapter 23.

5 HINTS ON DIAGNOSIS

It goes without saying that where facilities for X-rays, electroencephalograms, and brain scans are available they will, when appropriate, be brought to bear for diagnosing virus diseases of the CNS. The virus laboratory can also make a useful contribution, provided that there is good liaison with the clinicians.

In the first instance, you will need to take:

◆   throat swabs;

◆   faecal samples (don't forget the enteroviruses!);

◆   **paired** CSF and serum samples;

◆   fluid from any vesicular rash present.

If a brain biopsy is done, close collaboration between ward and laboratory is essential in order to extract the maximum information from the material.

Diagnostic procedures are mostly dealt with in Chapter 27, but it is appropriate here to describe the principles of serological diagnosis as applied to the CNS.

As mentioned in Section 1, evidence of the manufacture of specific antibody within the CNS can give a useful diagnostic lead. Immunoglobulins cannot normally gain access to the CNS compartment but, if permeability is increased as a result of inflammation, some may leak into the CSF. There are several ways of distinguishing between synthesis and leakage.

A simple method is to see whether the serum contains a workable titre of a reference antibody and then to determine whether the serum : CSF ratio of this antibody differs from that of antibody to the suspected virus.

Example. In a patient with suspected herpes simplex encephalitis, the titre of anti-HSV antibody was 512 in the serum and 32 in the CSF, i.e. a ratio of 512:32 = 16. Many people have measles antibody in their blood and this patient happened to have it at a titre of 128. The titre in the CSF however was only 2, giving a serum : CSF ratio of 64. In other words, there was four times more HSV antibody in the CSF than could be explained by non-specific leakage, thus helping to confirm the diagnosis.

Obviously, this method cannot be used if there is no suitable reference antibody in the serum. To avoid this difficulty, another method compares the serum : CSF ratio of the test antibody (in this instance anti-HSV) with that of albumin.

Using the figures from this example,

HSV antibody titre in CSF	$= 32$
HSV antibody titre in serum	$= 512$
CSF : serum antibody ratio (A)	$= 32/512 = 0.06$
CSF albumin concentration	$= 0.4$ g/l
Serum albumin concentration	$= 30.0$ g/l
CSF : serum albumin ratio (B)	$= 0.4/30.0 = 0.01$
$A/B = 0.06/0.01$	$= 6.$

The normal range is $0.05-1.46$, so that again the result indicates a higher titre of HSV antibody in the CSF than could be accounted for by leakage of protein.

Yet another variant, less specific but in some ways more informative, is the index described by Link, which compares the concentrations of both albumin and IgG in the serum and CSF.

All this emphasizes the importance of getting paired samples of blood and CSF; one without the other is of much less value.

At present, serological tests of this sort have the disadvantage that they cannot give results until the second week of illness, but the development of more sensitive tests for antibody and antigen that can be used routinely should decrease this diagnostic time-lag.

6 REMINDERS

◆ Many viruses cause diseases of the CNS, which can be classified in three groups:

1. **acute infections;**
2. **acute postexposure syndromes;**
3. **chronic diseases.**

◆ Virus can usually be demonstrated in the CNS in the acute infections, but not in the postexposure syndromes, in which **demyelination** is usually a prominent feature and which may result from an **abnormal immune response** to live or inactivated virus.

◆ The chronic infections are all rare, progressive, and fatal. They include: **subacute sclerosing panencephalitis**, a late result of measles infection; **progressive multifocal leucoencephalopathy (PML)**, a polyomavirus infection seen in immunodeficient patients; **Creutzfeldt-Jakob disease**, a spongiform encephalopathy seen mostly in elderly patients; and **kuru**, a similar type of infection formerly endemic in a New Guinea tribe and spread by contact with infected human brains.

◆   The **laboratory diagnosis** of virus diseases of the central nervous system sometimes demands special techniques, including those for **detecting viruses or their antigens**, and for determining whether **intrathecal synthesis of antibody** is taking place.

CHAPTER TWENTY-FIVE

INTRAUTERINE AND PERINATAL INFECTIONS

1 INTRODUCTION

Of the distressing situations encountered in medical practice, the loss of a wanted pregnancy or the birth of a malformed infant are among the most poignant; the latter may indeed carry the prospect of many years of misery and hardship for both child and parents. Some of these tragedies are caused by infection with viruses, bacteria, and even protozoa. It must, however, be stressed that **microbial infection is responsible for only a small proportion of the overall total of miscarriages and damaged infants**; in the UK, the incidence of birth defects is about 20 per 1000 live births, or 2 per cent. This background level must be kept in mind before drawing overly hasty conclusions about a causal relationship between an infection in pregnancy and an unfavourable outcome. Some virus infections other than those described in this chapter have been suggested as causes of fetal damage, but we feel that the evidence is not strong enough to warrant their inclusion.

In this chapter the following terms are used:

◆ **intrapartum**: during birth;
◆ **perinatal**: the period from a week before to a week after delivery.

2 PATHOGENESIS

The pathogenesis of fetal infections is beset by uncertainties, both because of the obvious difficulty of investigating them in humans and because, more

Table 25.1 Modes of intrauterine and perinatal viral infections

Virus	Transplacental	During birth	Shortly after birth
Rubella	++	–	–
Cytomegalovirus	+	++	++ (BM)
Herpes simplex	+	++	+
Varicella-zoster	++	+	+
Parvovirus	++	–	–
Enteroviruses	+ (Late)	++	++
Human immunodeficiency virus	+	++	+ (BM)
Hepatitis B	+	++	++
Human papillomaviruses	–	++	–

++, Most frequent route; +, less frequent route; (BM), can be transmitted via breast milk.

Table 25.2 Possible adverse outcomes of intrauterine and perinatal viral infections

Virus	Death of embryo/fetus	Clinically apparent disease at or soon after birth	Long-term persistence of infection, with or without clinical signs
Rubella	+	+	+
Cytomegalovirus	?	+	+
Herpes simplex	+	+	+
Varicella-zoster	+	+	+
Parvovirus	+	–	–
Enteroviruses	+	–	–
Human immunodeficiency virus	?	+	+
Hepatitis B	–	–	+
Human papillomaviruses	–	–	+

? = uncertain

often than not, there is no suitable animal model. Much of what is written is thus based on inference. In this chapter we shall concentrate as far as possible on what is reasonably certain.

There are two main routes by which intrauterine and intrapartum infections are transmitted. **Transplacental infection** results from a maternal viraemia and may take place at any time during pregnancy, the outcome depending on the virus concerned. There is little or no evidence that **ascending infection** from the cervix or vagina can penetrate the fetal membranes during pregnancy, but contact infection from these sources can certainly take place during delivery and is facilitated by a long interval between rupture of the membranes and birth. Table 25.1 shows the ways in which the viruses most frequently implicated in these infections gain access to the embryo or fetus.

Some of the wide range of effects exerted by viruses on the fetus are shown in Table 25.2. They may be **teratogenic**, resulting from a cytopathic effect during organogenesis in the first trimester of pregnancy, or from virus-mediated damage to organs that have completed their development. In practice, the distinction is often difficult; rubella and cytomegalovirus (CMV) are certainly teratogenic but others may or may not act in this way.

3 FETAL IMMUNITY

The first class of immunoglobulin to be synthesized by the fetus is **IgM**, which becomes detectable at about 30 weeks of gestation and attains about 10 per cent of the adult concentration by the time of birth. Being a comparatively large molecule, it does not cross the placenta from the maternal circulation; this is also true of **IgA**, which is not made by the fetus in significant amounts. The fetus starts to make **IgG** at 17 weeks but, because its cellular immune system is immature, it cannot make specific antibody in response to an infection. Its only protection—and a very important one—is maternal IgG that reaches it across the placenta. This immunoglobulin, of course, contains a full complement of the mother's antibodies, and we shall see later in the chapter how it comes into play.

At birth, the infant's own cell-mediated and antibody responses are relatively ineffective and interferon production is poor; however, the maternal antibodies in its circulation are—if the baby is breast-fed—supplemented by **IgA antibodies in the colostrum**. These are not absorbed into the bloodstream, but protect the gut against pathogens that gain access by the oral route, a particularly valuable property in areas where hygiene is poor.

4 SPECIFIC INFECTIONS

4.1 *Rubella*

We mentioned in Chapter 7 the discovery by Gregg in Australia, 50 years ago, of the effect on the fetus of rubella acquired in pregnancy. This was the first recognition that fetuses can be damaged by viruses.

Congenital rubella is the result of a **primary** infection of the mother (Chapter 18) during the first 16 weeks of pregnancy. In the absence of maternal antibody, the virus readily crosses the placenta and the time at which this

Table 25.3 Most frequent physical signs of severe congenital rubella, cytomegalovirus infection, herpes simplex and toxoplasmosis*

Defects	Expanded rubella syndrome	Cytomegalic inclusion disease	Generalized herpes simplex	Toxoplasmosis
Apparent at birth				
Hepatosplenomegaly, jaundice	+	+	+	+
Thrombocytopenia, petechial haemorrhages, purpura	+	+	+	+
Skin vesicles			+	
Microcephaly	+	+	+	+
Intracranial calcifications		+		+
Pneumonitis	+	+	+	+
Cataracts	+			+
Glaucoma	+			
Choroidoretinitis	+	+		+
Patent ductus arteriosus, lesions of pulmonary artery and aorta	+			
Bone defects	+			+
Apparent months or years after birth				
Sensorineural deafness, speech defects, mental retardation	+	+		+
Diabetes mellitus	+			

* All these infections, when severe, involve many body systems. A blank does not necessarily mean that the relevant defect is *never* present. The table aims to show only the most important associations.

happens is an important factor in determining the outcome. By and large, the earlier in gestation infection takes place, the worse the result; during the first month, the infection rate approaches 100 per cent.

Table 25.3 lists the most frequently seen abnormalities in full-blown congenital rubella — the 'expanded rubella syndrome' — and some other serious infections. Many systems are affected and the prognosis is very poor. Figure 25.1 gives an idea of the relative frequency of some of the major lesions; the chart is compiled from figures obtained during a major epidemic of rubella in New York during 1964, well before the introduction of rubella vaccine. It shows, for example, that, of 106 infants born after their mothers contracted rubella during the second month of pregnancy, only five (4.7 per cent) were clinically normal; 58 per cent had defects of the cardiovascular or central nervous systems, 29 per cent had eye defects (Fig. 25.2) and 72 per cent were later shown to have sensorineural deafness. Clearly, the inner ear is susceptible to damage over a longer period than the cardiovascular system or CNS. It is interesting that the curve for clinically normal infants is a mirror image of that for deafness, implying that aural damage is the clinical lesion best correlated with infection *in utero*.

Latent defects

Some infants who appear normal at birth are later discovered to have hearing defects, with or without some degree of mental retardation. Prenatal infection

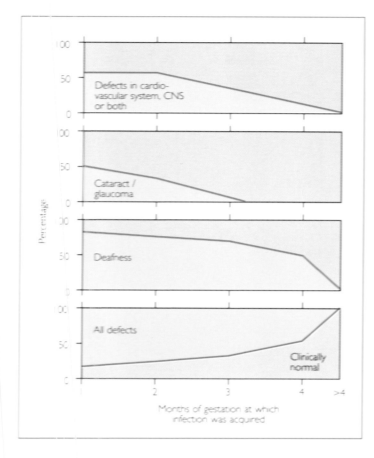

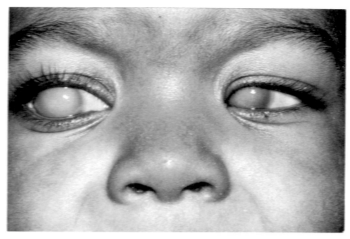

Fig. 25.1 Congenital defects in rubella-infected neonates (*left*). Data simplified from Cooper, Z. Ziring, P.R. *et al*. (1969), *American Journal of Diseases of Childhood* **118**, 18. The incidences of cardiovascular and CNS defects at the various stages of gestation were almost identical, and have been combined.

Fig. 25.2 Congenital rubella syndrome: bilateral cataracts in a 5-month-old baby (*right*). (Courtesy of the late Dr W. Marshall.)

often results in prolonged excretion of virus from the throat, a source of infection for susceptible contacts. Progressive subacute sclerosing encephalitis is a late and fortunately very rare complication.

Preconceptual rubella

A recent Anglo-German study showed no evidence of intrauterine infection in women who had rubella rashes before, or up to 11 days after, their last menstrual period.

Rubella immunization during pregnancy

There have been many instances of inadvertent vaccination against rubella during early pregnancy, followed by termination for fear of possible fetal damage by the live vaccine virus (Fig. 25.3). Although vaccine virus has been isolated from aborted products of conception, there is no evidence, from such pregnancies allowed to go to term, of rubella-associated defects, and termination should not be recommended on this ground. The figure also shows the

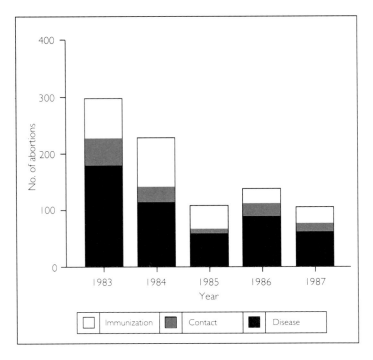

Fig. 25.3 Legal abortions following rubella in pregnancy. (Drawn from data in Table 29 in OPCS Series AB No.14: Abortion Statistics, England and Wales, 1987, by permission of the Office of Population Censuses and Surveys.)

diminishing rate of abortions undertaken because of rubella in pregnancy, or fear of its consequences.

Incidence of congenital rubella

Figure 25.4 shows that, following the introduction of immunization in 1970, the incidence of congenital rubella in the USA is now very small; nevertheless, continued vigilance is needed to prevent these disastrous infections.

Diagnosis of the mother

Because the clinical diagnosis of rubella is not always easy, laboratory tests are essential as a guide to the important decisions that must be made following a history of contact with rubella, or of a rubella-like illness, within the first trimester of pregnancy. Tests for **specific antirubella IgM and IgG antibodies** are made (Chapters 18 and 27), and action is taken according to the scheme shown in Fig. 25.5. **These tests must be done irrespective of a history of past immunization, which is not always reliable**.

The plan looks a little complicated, but is in fact quite logical; it is based on:

◆ the time course of the appearance of IgM and IgG antibodies (Fig. 18.2);

◆ the possible results of tests, arranged according to whether they are first done during, or outside the extreme ranges of the incubation period of

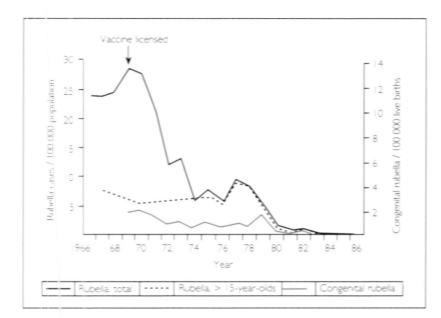

Fig. 25.4 Effect of immunization on incidence of congenital rubella. (Adapted from data in Morbidity and Mortality Weekly Report (1988). *Morbidity and Mortality Weekly Report* **36** 664).

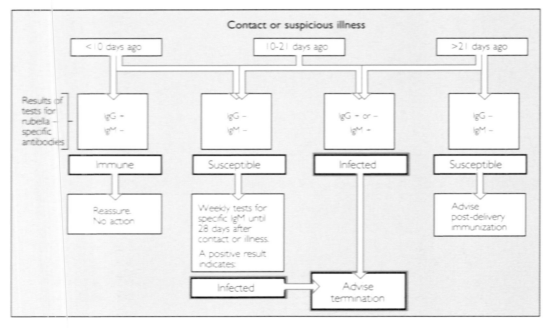

Fig. 25.5 Action following exposure to rubella in early pregnancy.

rubella (10–21 days). There is one trap to avoid. In addition to a known contact, say 14 days previously, there may have been another unsuspected contact some days later; if this seems a possibility, weekly tests for IgM antibody should be continued until the end of the longest possible incubation period has passed.

Diagnosis of the neonate

Virus isolation is used only for:

◆ confirming a diagnosis of prenatal rubella by testing the tissues of an aborted fetus;

◆ monitoring the shedding of virus from a congenitally infected baby.

Specimens are inoculated into Vero or RK 13 cells. Provided the culture conditions are properly controlled, a cytopathic effect is seen several days later; its specificity should be confirmed by immunofluoresence staining or by further passage in cell cultures.

The laboratory diagnosis of congenital rubella is confirmed by demonstrating **specific IgM antibody** in the cord or peripheral blood.

4.2 Cytomegalovirus (CMV)

The epidemiology and pathogenesis of fetal rubella are relatively straightforward. The behaviour of CMV is much more complex and there are many gaps in our understanding of intrauterine infections.

Risk of maternal infection

CMV infection is widespread, its prevalence being highest where living standards are low (Chapter 15, Section 5.4). On average, one might expect 60–70 per cent of women of childbearing age to have serological evidence of infection. Again on average, 1–2 per cent of seronegative pregnant women have primary infections during their pregnancy. An ill-defined proportion of people undergo periodic reactivations, and something like 10 per cent of pregnant women excrete CMV in their urine at some time during pregnancy.

Risk of fetal infection

Unlike rubella, transmission of CMV to the fetus takes place even in strongly seropositive women, perhaps because the virus travels within leukocytes and is to some extent sheltered from the effects of antibody. As might be expected, the effects on the fetus are usually more serious if transmission takes place during a primary infection, but it is well known that an infected woman can give birth to damaged infants in successive pregnancies. As might be expected, the earlier in gestation that transmission occurs, the worse the outcome for the fetus; those infected late in pregnancy tend to be born with inapparent infections.

Figure 25.6 shows that, of fetuses infected *in utero*:

◆ About 1 per cent die at or soon after birth with severe congenital defects.

◆ Four per cent have severe cytomegalic disease.

◆ Fifteen per cent appear normal but hearing defects and perhaps some degree of mental retardation become apparent later.

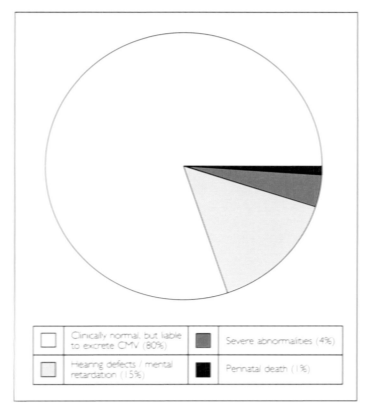

Fig. 25.6 Clinical outcomes of cytomegalovirus infection *in utero*.

☐	Clinically normal, but liable to excrete CMV (80%)	◼	Severe abnormalities (4%)
☐	Hearing defects / mental retardation (15%)	◼	Perinatal death (1%)

◆ The remaining 80 per cent are clinically normal but are liable to be persistently infected.

It has been estimated that, of 700 000 live infants born annually in the UK, about 2000 are congenitally infected with CMV; of these, approximately 200 are significantly damaged, making this infection a much greater hazard than congenital rubella now is.

Cytomegalic inclusion disease

This term is applied to the syndrome seen in infants born with the severe defects characteristic of CMV infection acquired early in pregnancy. It is a devastating generalized infection, liable to affect every system; Table 25.3 shows the most frequent abnormalities, but cardiovascular, gastrointestinal, and muscular defects also occur. The prognosis is poor, and some infants die shortly after birth.

Latent defects

In a small proportion of infected infants who appear normal at birth, hearing defects become apparent in early childhood; they may be associated with a degree of mental retardation.

Inapparent infection

The great majority of infected neonates (Fig. 25.5), notably those infected in the perinatal period, show no signs of infection, but continue to excrete virus in their saliva and urine for months or even years. They are potent sources of infection for susceptible people in close contact with them, particularly other infants in nursery schools.

Diagnosis

Clinically, cytomegalic inclusion disease must be distinguished from congenital rubella and toxoplasmosis, with both of which — particularly toxoplasmosis — it has many features in common (Table 25.3). In rubella, central cataracts and cardiovascular defects are particularly noteworthy, whereas toxoplasmosis is characterized by the high frequency of choroidoretinitis, hydrocephalus, and intracranial calcification. Congenital syphilis, now rare in the UK, also enters the differential diagnosis.

The laboratory diagnosis of CMV disease is made by finding the characteristic **'owl's eye' inclusions** (see Figure 15.7) in various organs at autopsy, or **specific IgM antibody** in cord or peripheral blood. Virus can be detected in cell culture (see Chapter 15, Section 5.5).

Prevention

Since the vast majority of CMV infections in adults are silent, there is no practicable way of knowing if or when a susceptible woman becomes infected while pregnant, and, even if there were, the position is complicated by the liability of recurrent infections to be transmitted to the fetus, and by uncertainty as to the outcome if this does happen. Under these circumstances, advice as to termination of pregnancy is fraught with difficulty.

Is it possible to immunize against CMV? Results with an experimental live attenuated vaccine have been promising, and there are those who advocate routine immunization of susceptible females. Nevertheless, there are uncertainties about the advisability of large-scale immunization with a herpesvirus vaccine, so far an unknown quantity in humans.

4.3 Herpes simplex viruses (HSV)

Neonatal infections are almost always caused by HSV-2 and are usually acquired during passage through an infected birth canal, although there are reports of a few infants born with severe defects who must have become infected *in utero*. Signs of intrapartum infection become apparent within a few days of birth; they range from a few trivial skin vesicles to massive and mostly fatal infections involving the brain, liver (Fig. 25.7), and other viscera. In these infants, the clinical picture is similar to those of rubella and CMV (Table 25.3), but is characterized by vesicles in which herpes virions can readily be detected. The incidence of such massive infections seems to vary

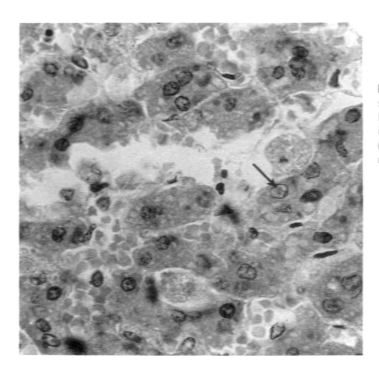

Fig. 25.7 Generalized congenital herpes simplex. Section of liver showing damaged hepatocytes, some of which contain intranuclear herpes inclusions. The clear halo (arrowed) surrounding the inclusion is characteristic.

from country to country. Neonatal herpes is diagnosed more frequently in the USA than in the UK; there is argument as to whether the difference is real, or due to failure to recognize some cases of herpetic meningitis (which seems unlikely). Be that as it may, the small number of cases diagnosed overall bears no relation to the number of women who must be infected. This may well be because mothers who have been infected for some time possess IgG antibody that crosses the placenta and protects the infant at birth. One would thus anticipate danger if a mother acquires a **primary** genital infection (Chapter 15) too late in pregnancy to have formed antibody by the time of delivery. Recognition of this situation demands Caesarean section. Diagnosis of a **recurrent** infection in pregnancy is not regarded by many British obstetricians as an absolute indication for section, although it almost always is in the USA.

Diagnosis and management

■■■■■ **Pregnant women** who undergo a primary infection with genital herpes near the time of delivery should be offered Caesarean section. Those with a past history should be swabbed every week from the 36th week until term to detect a recurrence; Caesarean section should be considered in the case of a positive result near the time of delivery but is not mandatory. **Infants** born to mothers with known HSV infection should be checked carefully during the first week, and swabbed from the nose, mouth, and skin. The appearance of herpetic vesicles dictates immediate treatment with acyclovir (Chapters 15 and 29).

The TORCH screen

Clinicians are fond of requesting this group of laboratory tests on neonates in whom there is any suspicion of a congenital infection with the viruses discussed so far. TORCH is an acronym for *to*xoplasmosis, *r*ubella, *c*ytomegalovirus, and *h*erpes simplex. A glance at Table 25.3 shows how many features these infections have in common, but such blunderbuss requests are not popular with laboratory staff, since a careful history and clinical examination often rules out one or more of these infections, making some or all of the tests redundant.

4.4   Varicella-zoster (VZV)

Of the herpesviruses that infect the fetus, VZV is fortunately the most rare; fortunately, because chickenpox in the mother during the first half of pregnancy may result in the birth of an infant with **severe scarring of the skin** — a particularly characteristic feature — hydrocephalus, and other malformations. More often, a mother who acquires varicella around the time of delivery transmits it to her infant *in utero*; if birth takes place less than 5 days after the maternal rash, the infant is liable to have a severe infection, with a fatality rate of about 30 per cent. If the interval is 5 days or more, the infection is milder, probably because the mother has developed antibody and passed it to the fetus. Occasionally, latent infection may be established, resulting in shingles a year or two later.

Diagnosis and management

Diagnosis is based on the distinctive vesicles, in which herpesvirus virions can be found by electron microscopy, and by the appearance in the maternal and neonatal blood of specific IgM antibody. Infants who have not yet developed a rash may be given prophylactic zoster immune globulin (ZIG, Chapters 15 and 28), or acyclovir if a rash does appear.

In hospitals, appropriate precautions must be taken to prevent cross-infection to susceptible patients and staff.

4.5   Parvovirus

Although parvovirus B19 can cross the placenta, the evidence to date does not suggest that it is teratogenic in the same way as rubella or cytomegalovirus. A prospective study by the Public Health Laboratory Service so far suggests a small excess risk of spontaneous abortion, and there are a few reports of fatal hydrops fetalis following infection later in gestation. There is thus no indication for advising termination. Needless to say, however, women with signs of parvovirus infection who know or suspect that they are pregnant must be rigorously tested to exclude rubella, which may be circulating in the community at the same time and has a similar incubation period.

4.6   *Enteroviruses*

Polioviruses, coxsackieviruses, and echoviruses are able to cross the placenta in late pregnancy, but the more usual mode of infection is during birth.

Poliovirus infection during pregnancy is now very rare in well immunized populations. The virus can cross the placenta and may cause abortion. The results of infection at term range from mild febrile illness to neurological deficits similar to those seen in later life. Although live poliovaccine is not advised in pregnancy, there is no evidence that it actually damages the fetus; termination should not be suggested on the ground of inadvertent vaccination.

Group A coxsackieviruses rarely affect the newborn. **Coxsackie B viruses**, on the other hand, may cause severe myocarditis, encephalitis, or a sepsis-like illness which must be distinguished from a bacterial infection.

Echoviruses usually give rise to a respiratory or gastrointestinal disease. Severe multisystem disease, affecting particularly the CNS and liver, has also been reported; two such outbreaks in the UK were due to echovirus 11.

4.7   *Human immunodeficiency virus (HIV)*

The decade since the start of the AIDS pandemic in 1981 has seen an accumulation of reports of unfortunate infants infected before, during, or shortly after birth by seropositive mothers, not all of whom had overt infection. We still lack detailed knowledge of modes of transmission and pathogenesis; so far, it seems that up to 50 per cent of infected mothers transmit HIV to their infants at birth, the outcome varying from a silent infection to full-blown AIDS within the next 2 years. HIV can cross the placenta in early pregnancy, but the main mode of transmission seems to be during labour, or by breast-feeding.

The Public Health Laboratory Health Service stated that to the end of October 1991 there were reports of 370 HIV-infected children in the UK, of whom 76 had AIDS; this number is steadily rising.

Management

Women known to be seropositive for HIV should not become pregnant, but, if they do, there is a strong case for termination. There is some evidence that Caesarean section reduces the risk of intrapartum transmission, but this needs confirmation.

4.8   *Hepatitis B*

The perinatal transmission of HBV plays a major part in perpetuating the infection in highly endemic areas, and is therefore described in Chapter 20.

4.9   *Human papillomaviruses (HPV)*

HPV-6 and 11 may be transmitted to the neonate during delivery, and later give rise to benign laryngeal papillomas (Chapter 21).

5   REMINDERS

◆ A number of viruses can seriously damage the fetus or neonate; with some, transmission via the **placenta** is the usual mode of infection, whereas others predominantly infect the baby during or shortly after delivery (**perinatal infection**).

◆ Some viruses — notably rubella and CMV — infecting the mother during early pregnancy are **teratogenic**, i.e. they interfere with organogenesis; both these and others can also damage organs and tissues that are already formed.

◆ Because the immune system of the fetus and neonate is immature, **maternal IgG antibodies** transported across the placenta and **IgA antibodies** in colostrum and breast milk are important protective factors.

◆ The adverse effects of most intrauterine viral infections include **fetal death**, **severe disease in the neonate** affecting a number of body systems, and **persistent postnatal infections**, with or without overt illness. Persistent but silent antigenaemia is an important aspect of perinatal HBV infection transmitted by carrier mothers.

◆ **Primary** infections of the mother with **rubella**, **CMV**, and **HSV** are more likely to damage the fetus than are reinfections or reactivations.

◆ The finding of specific **IgM antibody** is used to diagnose recent rubella in the mother, and, because it does not cross the placenta, **IgM antibodies in the cord or peripheral blood** are good indicators of infection of the neonate by rubella and other viruses.

CHAPTER TWENTY-SIX

VIRAL INFECTIONS IN PATIENTS WITH DEFECTIVE IMMUNITY

1 INTRODUCTION

In the preceding chapters there are many examples of the damage that can be wrought in otherwise healthy people by viruses despite the formidable battery of immune responses normally mounted against them. Much worse then is the plight of those unfortunates who lack the ability to respond adequately to microbial invasion. The problem is compounded by the fact that some viruses themselves impair immunity by damaging the cells that mediate immune responses. To put it briefly.

IMPAIRED IMMUNITY ⇌ SOME VIRUS INFECTIONS.

There are many types of immune defect, resulting in infection with all sorts of microbes, but here we must concentrate on those relating to viruses. The immunodeficiences fall under the following main headings.

◆ **Primary**, or congenital.
◆ **Acquired**:
  (1) **secondary to other diseases** (including some virus infections);
  (2) **secondary to various treatments** for other conditions ('iatrogenic').

Of these, the acquired or secondary forms are much more often encountered than the congenital varieties.

2 PRIMARY IMMUNODEFICIENCIES

Although these syndromes are comparatively rare, more than 20 varieties are listed. For practical purposes, however, we can classify them in two main groups (Table 26.1): those **predominantly involving B cells and hence immunoglobulin and antibody production**; and others **mainly affecting T cells and hence cell-mediated immunity (CMI)**. But, as we showed in Chapter 5, there is an intimate relationship between these two major arms of the immune system and in a number of syndromes both are involved to varying degrees. We shall here consider only those in which virus infections play a significant part.

2.1 *Deficiencies predominantly affecting B cells and antibody production* (Table 26.1)

Agammaglobulinaemia (or more often hypogammaglobulinaemia, since small amounts of immunoglobulin are usually detectable) presents between the ages of 6 and 24 months, at a time when the infant has lost its complement of maternal antibodies; its abnormal susceptibility to infection, usually bacterial but sometimes viral, is often the alarm that triggers investigation of its immune system. One variety is X-linked, affecting only males; a family history of severe infections in male infants is a good diagnostic pointer. More rarely, the trait is autosomal recessive. The defect is a **failure of maturation of B cells**; T cell function is normal. Replacement therapy with intravenous immunoglobulin is useful.

The **common variable immunodeficiencies** are also characterized by defective production of immunoglobulin but, as the name implies, the faults in lymphocyte production are diverse, involving B cells and T cell subsets in varying degrees. These syndromes are **late in onset**, usually presenting in young adults of both sexes; they are often familial. The patients often suffer from bacterial infections of the respiratory tract, intestinal giardiasis, and pernicious anaemia.

There is also a variety of **selective immunoglobulin deficiencies**, most often affecting production of IgA; people with this defect are usually healthy, although some of them get frequent respiratory infections. Defects in IgM and IgG production have also been reported.

2.2 *Deficiencies predominantly affecting T cells and cell-mediated immunity* (Table 26.1)

Severe combined immunodeficiency syndromes occur in young children and may be X-linked, or autosomal recessive, affecting both sexes. They carry

Table 26.1 Primary immunodeficiencies predisposing to virus infections

Predominantly affecting B cells and antibody production	Predominantly affecting T cells and cell-mediated immunity
Agammaglobulinaemia or hypogammaglobulinaemia (X-linked or rarely, autosomal recessive trait)	Severe combined immunodeficiencies
Common variable immunodeficiencies (late onset)	Thymic aplasia (DiGeorge's syndrome) Purine nucleoside phosphorylase deficiency
Selective immunoglobulin deficiencies	Lymphoproliferative syndrome with unusual response to EB virus Interferon deficiencies (alpha and gamma)

a high mortality, death being frequently due to fulminating bacterial or virus infections. They are associated with abnormalities of the thymus and thus of T cell development, but antibody production is also defective.

Thymic aplasia due to a developmental failure of the 3rd and 4th pouches (**DiGeorge's syndrome**) is not so serious; these children too may be predisposed to virus infections, particularly with cytomegalovirus (CMV), but in many instances there is spontaneous recovery of T cell production. They often respond well to fetal thymus and bone marrow grafting.

Purine nucleoside phosphorylase deficiency is a rare condition resulting from a structural defect in chromosome 14; the end result of absence of the enzyme is impairment of T cell function and these children often die from overwhelming herpesvirus infection.

At least in the Western world, infection with **Epstein–Barr virus (EBV)** is not normally dangerous. There is, however, a rare T cell defect, usually but not invariably X-linked, which impairs the immune response to this virus; such patients develop a range of potentially fatal syndromes including severe mononucleosis, B cell lymphomas, bone marrow aplasia, and agammaglobulinaemia.

It has recently been recognized that in some people, possibly several per cent of the population, there are **defects in production of alpha or gamma interferon**, i.e. the varieties produced in T cells. There is some evidence that such persons are unusually prone to respiratory and herpesvirus infections; a more serious situation arises if they are unlucky enough to contract hepatitis B, since they are then liable to become chronic carriers (see Chapter 20).

2.3 *Virus infections associated with primary immunodeficiencies* (Table 26.2)

In Chapter 4, Section 3.2 , we mentioned briefly that the sort of cytopathogenic effect induced by a virus gives a clue to its behaviour in relation to the immune

Table 26.2 Virus infections associated with primary immunodeficiencies

Defect	Infections
B cells	Polioviruses (natural infection and live vaccine): paralysis Enteroviruses: encephalomyelitis, myositis Rotaviruses
T cells	Myxo- and paramyxoviruses Herpesviruses Papillomaviruses
Interferon	Respiratory viruses Herpesviruses

responses: lytic viruses (the 'bursters'), e.g. the enteroviruses, are influenced more by antibody, whereas the 'creepers', notably the enveloped agents such as herpes-, myxo-, and paramyxoviruses, are controlled more by CMI. Table 26.2 shows that the immune deficiency syndromes illustrate this point rather well.

B cell defects are associated particularly with **enterovirus** and occasionally **rotavirus** infections. Poliomyelitis is especially dangerous, with a high rate of paralysis; inadvertent immunization with the live vaccine is also liable to cause paralysis. According to the serotype, echoviruses may cause severe meningoencepalitis or myositis, and rotavirus infection may persist for long periods, with chronic diarrhoea and shedding of virus.

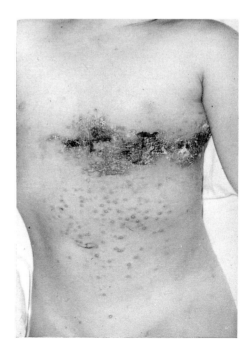

Fig. 26.1 Herpes zoster in immunosuppressed child. In addition to the characteristic dermatomal distribution there are vesicular lesions elsewhere on the chest and abdomen. (Courtesy of the late Dr W. Marshall.)

That impairment of T cell activity and CMI opens the way for the 'creeper' viruses was dramatically illustrated when smallpox vaccine was widely used; children with agammaglobulinaemia but with intact CMI responded normally, whereas those who also had defective T cell function developed generalized vaccinia which often proved fatal. In **measles**, impaired CMI may result in a life-threatening infection characterized by absence of rash and giant-cell pneumonitis. Nowadays, the main danger to such patients in Western countries is from chronic and occasionally generalized infections with **parainfluenza and influenza**; **herpesviruses**, particularly cytomegalovirus and varicella-zoster (Fig. 26.1) are also a threat,

A rare condition in which CMI is greatly depressed, epidermodysplasia verruciformis, is particularly associated with papillomavirus infections causing warts that in these patients may become malignant (see Chapter 21).

3 ACQUIRED IMMUNODEFICIENCIES SECONDARY TO OTHER DISEASES AND THEIR TREATMENT (TABLE 26.3)

Some **malignancies**, notably those of the blood and lymphoreticular system, are notorious for impairing the immune response and increasing susceptibility to virus infections. Paradoxically, this problem has in recent years been enhanced by the successful use of drug and radiation therapy given both for their direct cytotoxic effects on the tumour cells and as a preliminary to bone marrow transplantation for leukaemia; such treatments are profoundly immunosuppressive, a property that is also exploited to prevent rejection of transplanted solid organs such as the kidney.

Table 26.3 Acquired immunodeficiencies

Immunodeficiency secondary to

Other illnesses	Treatment
Malignancies, especially of blood and lymphoreticular system	Cytotoxic and immunosuppressive drugs and irradiation for tumour therapy or to prevent transplant rejection
Renal failure/dialysis AIDS Measles/malnutrition	

Table 26.4 Relative frequency and severity of virus infections in different immunodeficiency states

Virus infection	Immunodeficiency due to				
	Primary defects	Solid tumours	Haematological malignancies	Cytotoxic treatment	AIDS
Herpes simplex	+ +	+	+ +	+ +	+ +
Varicella	+ + +	+ +	+ + +	+ +	
Herpes zoster		+	+ +	+ +	+ +
Cytomegalovirus	+ +	+	+	+ + +	+ + +
Epstein–Barr virus	+ + +			+ +	
Myxo- and paramyxoviruses	+ +		+ +		
Adenoviruses				+ +	+ +
Enteroviruses	+ + +	r	r	r	
Rotaviruses	+ +				
Papovaviruses	+ +			+ +	+

Based in part on data from Wong, D.T. and Ogra, P.L. (1983). *Medical Clinics of North America*, **67**, 1075–93.
+ + +, very common and often severe; + +, common, moderately severe; +, infrequent or mild; r, rare.

Certain **virus infections** are themselves immunosuppressive. The prime example, and the one best understood, is that of the **human immuno-deficiency viruses (HIV)** which destroy T-helper cells and thus open the door to so-called opportunistic infections by a variety of microbes. As well as being a hazard to the immunocompromised patient, **CMV** is itself immunosuppressive and may facilitate reactivation of other herpesviruses. That infection with **measles virus** impairs CMI is well known and is neatly demonstrated by the failure of tuberculin to induce a delayed hypersensitivity skin reaction in positive subjects tested within 3 weeks of being given live measles vaccine. The opposite side of this coin is the observation that in African children malnutrition depresses CMI and thus delays recovery from measles.

Tabel 26.4 gives an idea of the relative frequency and severity of various virus infections in the main categories of immunodeficiency. Remember that, although such infections figure most prominently in the column headed 'Primary defects', these conditions are much more rare than those resulting from disease or therapy.

Types of infection

First and foremost it is the **T lymphocytes** that are suppressed, immediately suggesting that infections with herpes and other 'creeper' viruses will dominate the scene. That this is so is apparent from Table 26.4 and 26.5. Other viruses often infecting immunosuppressed patients include **adenoviruses**, which cause respiratory and enteric infections, and **papillomaviruses**, which may give rise to persistent warts. **Progressive multifocal leucoencephalopathy** is a demyelinating disease occasionally seen in older patients whose immunity

Table 26.5 Virus infections associated with malignancies, transplants, and related cytotoxic treatment

Virus	Remarks
Herpesviruses	
Herpes simplex	Severe local infections Encephalitis Occasional generalization
Chickenpox	In children ⎫
	⎬ **Pneumonitis**
Herpes zoster	In adults ⎭
Cytomegalovirus	Primary infection (from transplant); more often reactivation. **Pneumonitis**
Epstein–Barr virus	Occasional B–cell lymphoma, especially after renal allograft
Other viruses	
Adenoviruses	Gastrointestinal and respiratory infections
Myxo- and paramyxoviruses	Persistent influenza, parainfluenza, and respiratory syncytial virus infections
Papillomaviruses	Persistent warts
Polyomavirus (JC)	Progressive multifocal leucoencephalopathy in older immunosuppressed patients

is impaired by malignant disease or cytotoxic treatment. It is caused by the JC strain of **polyomavirus** and is uniformly fatal (Chapter 21).

4 SOME SPECIAL PROBLEMS (TABLE 26.5)

Many severe virus infections are associated with immunodeficiencies due to malignant disease and cytotoxic therapy, and are thus particularly likely to be encountered in hospital practice.

4.1 *Renal failure and long-term dialysis*

The antibody responses of these patients are intact but cell-mediated immunity is depressed. The main danger used to be hepatitis B, acquired from blood transfusions: the initial infection was often mild, but often developed into the carrier state, so that hepatitis B was endemic in dialysis units. Screening of blood donors and the institution of rigorous safety procedures has, for practical purposes, eliminated this problem in the UK and the main danger now for dialysis patients is **non-A non-B hepatitis**. Unfortunately, hepatitis B in dialysis units is still a problem in some countries. Circumstances occasionally make it necessary for a patient normally resident in the UK to be dialysed abroad; caution is needed when making such arrangements, first to protect the individual, who must be given hepatitis vaccine, and second to protect the home dialysis unit against the possibility of importation of infection when the patient is readmitted.

4.2   *Renal transplant patients*

In addition to the immunosuppression resulting from chronic renal failure, renal transplant patients are subject to other hazards.

◆ **Immunosuppressive drugs** such as azathioprine and cyclosporin A designed to prevent graft rejection enhance both susceptibility to new infections and reactivation of latent herpesviruses.

◆ **Cytomegalovirus (CMV) infection** presents special problems. From 50 to 80 per cent of adults are infected with this agent and the risk to transplant patients are twofold:

(1) reactivation of a latent infection;

(2) acquisition of the virus by a seronegative (i.e. non-immune) recipient from a seropositive (i.e. infected) donor either of transfused blood or of the kidney itself; in the case of blood, the virus appears to be transferred in the granulocytes.

Active CMV infection can be demonstrated in the great majority of these patients within the first 6 months after transplant. It may not cause overt illness and is then recognized only by laboratory tests. On the other hand, it may give rise to an episode of fever, glomerulonephritis with graft rejection, hepatitis, or, worst of all, pneumonitis, which carries a mortality of over 80 per cent and is the main cause of death in renal transplant recipients. Fortunately, this complication occurs only in a small minority of cases.

Prevention of CMV infection

Ideally, seronegative recipients of renal and other allografts should receive organs from seronegative donors, but this is not always feasible. Another approach now being tried is prior immunization of seronegative recipients with an experimental CMV vaccine; the evidence so far suggests that this procedure may not prevent subsequent CMV infections but diminishes their severity.

A substantial proportion of renal transplant patients start to excrete polyomavirus (usually the BK strain but sometimes JC) in their urine a few weeks after operation. These infections are usually reactivations; they are nearly always subclinical, but occasionally cause temporary depression of kidney function.

4.3   Malignancies of the blood and lymphoreticular systems

The immune responses of patients with malignant diseases of the blood and lymphoreticular systems are impaired even in the absence of treatment. Their susceptibility to infection is increased even more by the intense chemotherapy and irradiation preceding allogeneic **bone-marrow transplant (BMT)**, during which the cells of the immune system — B and T lymphocytes, macrophages, and other antigen-presenting cells — are virtually destroyed, to be replaced eventually by donor cells. This situation is obviously fraught with danger, and, particularly in view of the successes now being obtained in the treatment of children with acute lymphoblastic leukaemia, it is a tragedy to lose such a patient as a result of infection. Infections with herpesviruses, mostly due to reactivation, are a particular problem.

Herpes simplex virus (HSV) is usually the first to reactivate with a peak incidence at about 2 weeks. Dissemination is unusual, but in bone-marrow recipients the local lesions may be very extensive. If HSV infection threatens, it can be prevented or modified with acyclovir.

Epstein–Barr virus (EBV) reactivates in about 50 per cent of patients but does not usually cause significant illness, although EBV-associated lymphomas have been reported.

Primary infection with varicella-zoster virus (VZV) is a serious hazard in leukaemic children, whether or not they have had bone-marrow transplants. Reactivations cause herpes zoster in about 50 per cent of BMT recipients and may disseminate like chickenpox. **Pneumonitis** is a serious complication but can be treated with acyclovir.

Cytomegalovirus (CMV) infections pose the major threat to BMT patients, who are at greater risk than recipients of solid organs: **interstitial pneumonitis**, with a mortality rate of 80–90 per cent, occurs in about a third of them. Since the treatment of these infections with antiviral drugs is relatively ineffective, other measures must be employed.

◆ Frequent **monitoring for CMV reactivation by tests giving rapid results** is important, since timely modification of the immunosuppressive regime may help to reduce the severity of infection.

◆ Administration of **anti-CMV immunoglobulin** to seronegative patients before and after BMT has been claimed to reduce the infection rate.

◆ A recent experimental approach has shown promising results. Workers in the UK showed that immunization of donor marrow cells *in vitro* with tetanus toxoid or hepatitis B vaccine results in adoptive transfer to the recipient of B cell responses to these antigens. They then showed that the incidence of CMV pneumonitis was significantly reduced in **seropositive** recipients receiving marrow from **seropositive** donors. A possible explanation is that B stem cells from the donor, which are already primed to produce CMV antibody, are stimulated to do so when exposed to CMV antigen in the recipient.

Asymptomatic infections with polyomaviruses also occur, and are similar to those in renal transplant patients. (Section 4.2).

Contact with cases of measles or chickenpox is a particular hazard for leukaemic children who are not already immune since both viruses can give rise to giant-cell pneumonitis with high mortality rates (Figs. 26.2 and 26.3). VZV may also cause hepatitis or encephalitis. Immunization of such children with live vaccines is normally contra-indicated, but, in case of exposure to **measles**, prompt administration of normal immunglobulin (HIG) will abort infection or diminish its severity. A live vaccine against **chickenpox** has been developed in Japan and used successfully to prevent leukaemic children in hospital acquiring this infection. In view of the danger of using live virus vaccines in immunodeficient patients, it must have taken some courage to mount the first clinical trials but, fortunately, the vaccine seems to have no side-effects other than causing a very mild attack of varicella in a minority of patients.

4.4 *Solid tumours*

Table 26.4 shows that these patients, who tend to be older than the leukaemics, suffer mostly from **herpesvirus infections**, VZV often being particularly troublesome in the elderly.

Fig. 26.2 (*Below left*) Section of lung from measles pneumonitis. Note giant cell containing intranuclear (*I*) and cytoplasmic (*C*) inclusions (arrowed).

Fig. 26.3 (*Below right*) Section of lung from varicella-zoster pneumonitis. By contrast with the measles giant cell (Fig. 26.2) the inclusions (arrowed) are all intranuclear.

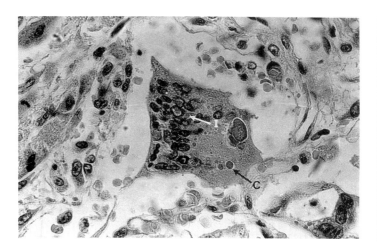

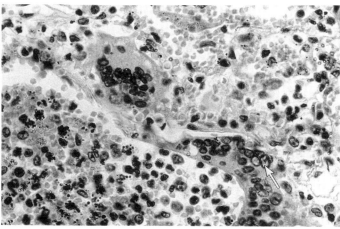

4.5   AIDS

This is the prime example of immunosuppression by a virus; HIV destroys the T helper cells and thus opens the way for so-called opportunistic infections by a wide variety of pathogens. As far as other viruses are concerned, **HSV eruptions** in the orofacial, perianal, and genital areas are common; **herpetic oesophagitis** also occurs. **Cytomegalovirus** is so constantly present in these patients that before the isolation of HIV it was suggested as a possible cause of the syndrome. It may cause **febrile illness, pneumonitis**, or **chorioretinitis**. Adenovirus and papovavirus infections also occur.

5   REMINDERS

◆ Patients with defects of immunity are **abnormally susceptible to infections with microbes, including viruses**; and **some viruses** (e.g. HIV, measles, CMV) can themselves **impair immunity by damaging the cells that mediate immune responses.**

◆ The **primary** (congenital) immunodeficiencies are comparatively rare. Those predominantly affecting **B-cell** function are mostly associated with infections by **cytolytic viruses**, e.g. enteroviruses, whereas patients with defects mainly of **T cell** functions are at most risk from **enveloped cell-associated** viruses, e.g. herpes and paramyxoviruses.

◆ **Acquired immunodeficiencies** are secondary to **malignant disease** and to **cytotoxic drugs** and **irradiation** given therapeutically or to prevent transplant rejection. They are met much more often than the primary immunodeficiencies.

◆ In these patients **T-cell function is impaired** so that the main risks are **reactivations of herpesviruses**, particularly CMV. Graft recipients who are not immune to CMV may acquire a primary infection from a seropositive donor.

◆ Cytotoxic treatment apart, children with **leukaemia** are at particular risk of primary infection with **measles** or **chickenpox**, both of which may cause serious **giant-cell pneumonitis**. They should receive passive protection with normal immunoglobulin if exposed to measles; a varicella-zoster vaccine is undergoing clinical trials with promising results.

PRACTICAL ASPECTS

CHAPTER TWENTY-SEVEN

THE LABORATORY DIAGNOSIS OF VIRAL INFECTIONS

1 INTRODUCTION

Many of the pathogenic bacteria can be isolated from clinical specimens and assessed for sensitivity to a wide range of antibiotics in a matter of hours. This is not yet true for viruses, isolation of which by traditional methods may take several days or, in extreme cases, weeks. In the virology laboratory, therefore, the emphasis is still on serological tests for specific antibodies, but the arrival of rapid and sensitive techniques for detecting viral antigens is now effecting a major improvement in the standard of service that can be offered to the clinician.

Among these methods, detection of viruses and their antibodies by automated methods such as enzyme-linked immunosorbent assay (ELISA) and the use in such tests of highly specific monoclonal antibodies, has revolution-ized the way in which clinical virology laboratories operate. The commercial availability of reagents, such as conjugates for immunofluorescence and test kits, e.g. for ELISA and latex agglutination tests, relieves the laboratory of the chore of preparing them and provides well standardized materials. They are, of course, relatively expensive; another factor, that cannot be overemphasized, is the danger of relying slavishly on the results of test kits without continual monitoring **within the individual laboratory** of their sensitivity and specificity. In various countries, including the UK, USA, and Australia, this task is helped by external quality assurance schemes whereby microbiology diagnostic laboratories are at intervals sent specimens from a central source and asked to report on the results of various tests. Scores are awarded both for correctness and speed of response, so that each laboratory has a continuing and impartial record of its performance.

In this chapter, we aim to outline the main methods in current use, concentrating much more on principles than on technical details. In describing virological diagnostic tests, it is usual to consider them under the headings of **identification of viruses (or their antigens)** and **detection of antibodies**, but, since modern methods have blurred the distinction between them, we shall consider first the more recently introduced techniques of **rapid diagnosis** and then those more traditional methods that still make a useful contribution.

2 SENDING SPECIMENS TO THE LABORATORY

Each laboratory has its own way of doing things, but the procedures outlined in this section provide a general guide.

To do its job properly, the laboratory must be provided with:

◆   the right specimens;

◆   taken at the right time;

◆   stored and transported in the right way.

Table 27.1 summarizes the types of specimen to be collected when viral infections of various body systems are suspected.

Table 27.1 Specimens required for isolation of virus or detection of antigen

Disease	Specimen
Respiratory infection	Nasal or throat swabs; postnasal washing
Gastrointestinal infection	Faeces (rectal swab not so satisfactory)
Vesicular rash	Vesicle fluid, throat swab, faeces
Hepatitis	Serum, faeces
Central nervous system	Cerebrospinal fluid, throat swab, faeces
AIDS	Unclotted blood

NB. In addition to the above, 5–10 ml of clotted blood for serological tests is always required.

Blood serum. If a syringe rather than a vacuum tube is used, remove needle before placing blood in container so as to avoid haemolysis.

Swabs from skin lesions or throat must be taken vigorously to collect adequate material, and squeezed into a tube of **transport medium**, usually tissue culture medium containing antibacterial and antifungal antibiotics to inhibit contaminants.

Vesicle fluid for electron microscopy is collected on the tip of a scalpel blade, or large needle, spread over a small area on an ordinary microscope slide, and allowed to dry.

Cerebrospinal fluid is collected in a dry sterile container.

Faeces, which should be placed in a dry sterile container, are preferable to rectal swabs for virus isolation.

Storage. Specimens should be placed in secure plastic bags and labelled in accordance with local practice. They should go to the laboratory as soon as possible after collection; if kept overnight, they should be held at 4°C.

Request forms. Much time will be saved if these are properly completed. Brief indications of the date of onset of illness, clinical signs, and suspected diagnosis are more important than a specification of the tests required.

3   RAPID DIAGNOSTIC METHODS

Many of these techniques depend upon serological reactions between viruses, or their antigens, with specific antibodies. It follows that tests of this type can often be used to identify either an unknown virus with an antibody of known specificity or, conversely, an unknown antibody with a known antigen.

3.1   *Monoclonal antibodies*

So-called specific antisera prepared by injecting animals with crude viral antigens are in fact a mixture of antibodies, many of which are irrelevant to the tests for which they are to be used. The specificity of tests involving antibodies as reagents has been greatly improved by the use of monoclonal antibodies (MAb), directed at specific viral antigens. In brief, mice are immunized with a particular virus; their spleen cells are then fused with a continuous line of mouse myeloma cells, which are capable of producing large quantities of antibody. A single fused cell making antibody of the required specificity is identified and cloned by limiting dilution. Since all the spleen B cells produce antibodies of different specificities, this process is very laborious, but, if successful, a cell line can be prepared from the single cloned cell that produces large amounts of antibody of a single specificity indefinitely.

3.2 *Immunofluorescence tests*

Principle

Direct fluorescent antibody method. Virus or viral antigen is reacted with a specific antiserum which is coupled with a fluorescent dye (fluorescein isothiocyanate, FITC). The dye becomes visible as a green fluorescence viewed by ultraviolet microscopy. The great disadvantage is that many specific sera must be labelled in order to test for a range of viruses.

Indirect fluorescent antibody (IFA) method. As for the direct method, except that the specific serum is unlabelled; instead, the dye is attached to a second serum prepared against globulins from the species in which the specific serum was made (Figs. 27.1, 27.2(a)). For example, antibodies to human immuno-globulins are often made in rabbits or goats. This is called a 'sandwich' method, for obvious reasons and has the great advantage that only one labelled (antis-pecies) serum is needed to test for many viruses.

Rather than FITC, the label may be immunoperoxidase, which is then reacted with a substrate to give a precipitate visible by ordinary light microscopy.

Applications

The method may be used to identify viral antigen in a clinical specimen, e.g. RSV in throat washings, or in a cell culture previously inoculated with the specimen. The latter method has been used with much success to detect early CMV antigens induced in cell cultures within 48 h of inoculation — a much faster method than waiting many days for a cytopathic effect.

Fig. 27.1 Influenza antigen in Vero cells stained by the indirect immunofluorescence method.

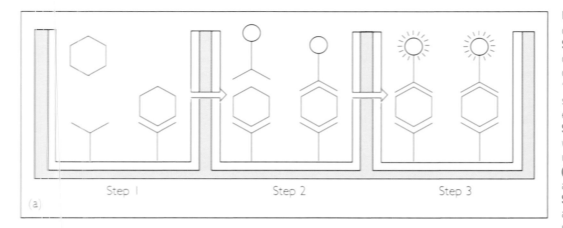

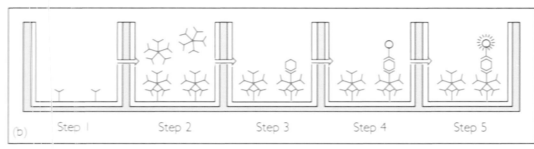

Fig. 27.2 (a) Direct identification of antigen by capture and ELISA. **Step 1**: Addition of specimen containing antigen which combines with the specific 'capture' antibody on a plastic surface. **Step 2**: Addition of enzyme-labelled specific antibody. **Step 3**: Substrate is added, reacts with bound enzyme, and undergoes colour change. (b) Identification of specific IgM antibody by capture and ELISA. **Step 1**: Plastic surface coated with antibody to IgM. **Step 2**: Patient's serum added; IgM molecules are captured by the anti-IgM. **Step 3**: After washing to remove unattached IgM, test antigen is added and combines with any captured IgM of the same specificity. **Steps 4 and 5**: as steps 2 and 3 in (a). Note that the captured IgM molecule on the left, having no specificity for the test antigen, does not react.

3.3 Enzyme-linked immunosorbent assay (ELISA) and radioimmunoassay (RIA)

ELISA (Fig. 27.2) is becoming the workhorse of many virology laboratories since it is automated, kits are commercially available, and it can be adapted to the identification of many viral antigens and antibodies such as the p24 antigen of HIV-1, hepatitis antigens, rotavirus, and the corresponding antibodies.

Principle

This is very similar to that of IFA. The main differences are as follows.

1. Instead of a fluorescent dye, the label is either an enzyme (ELISA) or radioactive iodine (RIA).

2. Specific binding of the labelled antibody (or antigen) is detected by reacting the enzyme with a substrate which then produces a visible colour in the reaction mixture (ELISA) or by counting radioactive emissions (RIA).

3. Rather than on a microscope slide, the reaction takes place in a tube or, more usually, a multiwell plastic plate, in which the reactions are read by

photometry (ELISA) or gamma counter (RIA) and printed out automatically. A very commonly used kit contains capture antibody on a bead.

Both ELISA and RIA can provide quantitative estimations of the amounts of antigen or antibody present. For various technical reasons, including safety, ELISA has, however, virtually replaced RIA in the routine laboratory. As with IFA tests, thorough washing between the various stages to get rid of unbound reagents and the use of positive and negative controls are essential.

3.4   *Latex agglutination tests*

Latex particles are coated with viral antigen and agglutinate when mixed on a slide with specific antiserum. The test is rapid, easy to read, and does not require complicated equipment. It is, however, liable to prozone effects, giving false negative results at low dilutions of serum.

3.5   *Electron microscopy (EM) and immunoelectron microscopy (IEM)*

Principle
Samples are negatively 'stained' with phosphotungstic acid, i.e. the virions, which are not penetrated by the stain, stand out as white particles on a dark background. At least 10^6 particles must be present on the EM grid to stand a chance of being identified; it is sometimes necessary to use concentration methods.

Applications

◆   Rapid identification of morphologically distinctive virions, usually directly in clinical specimens, but sometimes in cell culture fluid. HSV and VSV can be readily identified in vesicle fluid, although, being identical in appearance, they cannot be distinguished from each other by EM.

◆   Identification of viruses that cannot be grown in cell culture. These include rotaviruses, adenoviruses, and 'small round' viruses in faeces (Chapter 11) and HBV in blood.

◆   The value of the test may on occasion be increased by using IEM, which is the addition to the specimen of specific immune serum that agglutinates a particular virus, thus making the virions easier to find and adding serological specificity to their identification.

3.6 *Detection of viral genome by nucleic acid hybridization and the polymerase chain reaction*

These two highly sensitive methods are used to detect viral genetic information in infected cells. The latter method depends on a knowledge of the nucleotide sequence of the viral DNA or RNA genome, and at present, is used only in specialized laboratories, although its adoption for routine use is only a matter of time.

Principle of hybridization methods

Dot-blot hybridization involves extracting the nucleic acid — usually DNA — from the specimen and denaturing it into single strands. Spots of the DNA solution are placed on a nitrocellulose filter, and treated with a probe consisting of a labelled stretch of DNA or RNA complementary in sequence to the specific region being sought in the specimen. The label may be a fluorescent dye or a radioisotope. The principle of *in situ* hybridization is similar, except that the specific nucleic-acid sequences are labelled directly in tissue sections.

Applications

Viral nucleic acid hybridization techniques are very sensitive and can be used, for example, to detect the genomes of papillomaviruses and herpesviruses in tissues and enteric viruses in faeces.

Outline of the polymerase chain reaction (PCR)

Two distinct oligonucleotide primer sequences, one on each strand of the target DNA molecule (solid block, Fig. 27.3) are added to a clinical sample which has been treated with heat (94°) and detergent to denature the strands of viral DNA (Step 1). The primers specifically hybridize with the homologous nucleotide stretches on the viral DNA genome. A DNA polymerase (open square) called TAq polymerase (from *Thermophilus aquaticus*), which acts at high temperature, is added. After 1 min the temperature is reduced to 52°C for 20 s to allow annealing of primer (step 2) and the temperature is then raised to 72° for 5 min to allow DNA polymerization to occur (Step 3). Under these conditions, and only if the oligonucleotide primers have hybridized, the TAq enzyme generates multiple copies of the nucleotide stretch between the two primers. Multiple cycles of DNA denaturation, annealing of primers and polymerization can be programmed in the microprocessor-driven heating block. In this manner, a portion of a single molecule of viral DNA can be amplified a millionfold in a few hours to give a quantity of DNA that can be separated in a polyacrylamide gel and visualized by addition to the gel of ethidium bromide and exposure to UV light.

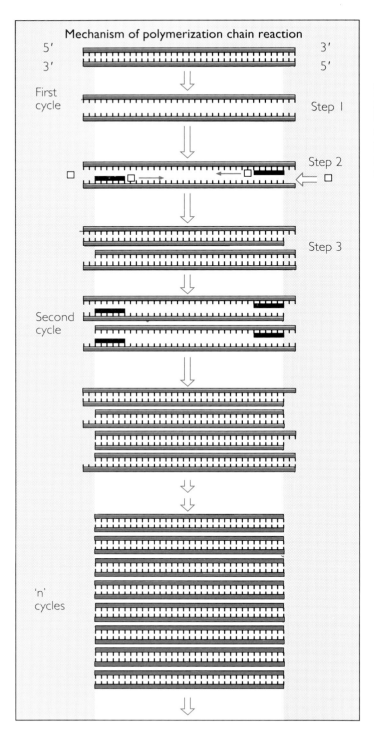

Fig. 27.3 Detection and amplification of viral genome by the polymerase chain reaction. **Step 1**: Denaturation (separation) of viral dsDNA. **Step 2**: Addition of oligonucleotide primer (■) and TAq polymerase (□). Annealing of primer. **Step 3**: DNA transcription and amplification.

Applications

The PCR method is used in specialized laboratories to detect HIV proviral DNA, cytomegalovirus DNA, and hepatitis B DNA in clinical samples. The method is exquisitely sensitive: the most important technical problem is the danger of false positives. For example, if a different laboratory in the same building has handled plasmid DNA of the same viral origin then contamination of buffers or equipment with a few molecules of plasmid DNA is difficult to avoid. This contaminating DNA could be amplified and, unless rigorous controls are incorporated, give a false positive result.

4 | VIRUS ISOLATION IN CELL CULTURES

Suitable cell cultures (Chapter 3) are inoculated with $100\,\mu l$ of the clinical specimen (throat washing, faeces, etc.) contained in transport medium (Section 2); the cell culture medium contains antibiotics to inhibit bacterial and fungal contaminants. The cultures are then observed for cytopathic effects (CPE, Fig. 27.4), which may appear within 48 h (e.g. enteroviruses, herpes simplex), or be delayed for as long as 14 days (e.g. cytomegalovirus). CPE fall into the following categories according to whether they are caused by viruses of the 'burster' (lytic) or 'creeper' types (Chapter 4).

◆ **'Burster' viruses** (e.g. enteroviruses): rounding up and lysis (Fig. 27.4(a));

◆ **'Creeper' viruses** (e.g. herpesviruses, paramyxoviruses): formation of multinucleate giant cells (syncytia), with or without 'ballooning' of clumps of cells (Fig. 27.4(b)–(d)).

A few viruses, although replicating in the cell culture, cause no visible CPE and are detected only by their ability to make the cells resistant to superinfection with a second virus. Other viruses not causing CPE (e.g. influenza virus) can be detected by immunofluorescence (Fig. 27.1) or by their capacity to bind red blood cells (haemadsorption). Some viruses, e.g. certain enteric viruses (Chapter 11) and hepatitis B (Chapter 20), cannot be grown in cell culture systems.

For isolating HIV-1 from AIDS patients a special technique had to be developed, since this agent grows only in human lymphocytes, which cannot normally be maintained in culture. This difficulty was overcome by stimulating them with a plant lectin, phytohaemagglutinin, and interleukin-2. The cytopathic effect of HIV-1 in such a culture is shown in Fig. 27.4(d).

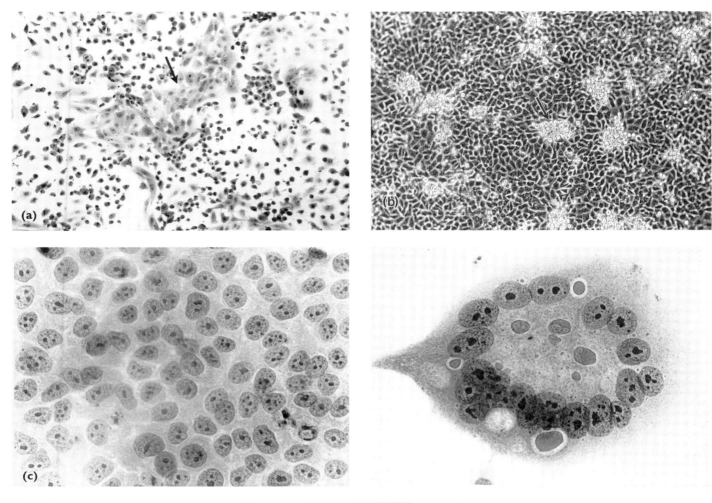

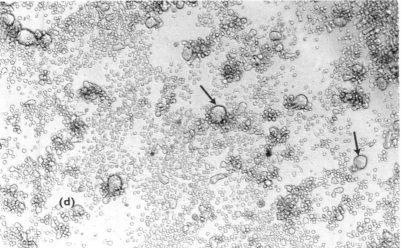

Fig. 27.4 Cytopathic effects (CPE) of viral infection. **(a) Enterovirus infection of a continuous line of human embryo lung cells (HEL)**. Areas of rounded, dead or dying cells lie between islands of normal cells (arrowed). **(b) Herpes simplex virus type 2 infection of baboon kidney cells** (phase contrast). The affected cells (arrowed) are swollen and refractile. **(c) High-power view of a continuous line of human epithelial cells** (left) **and a multinucleate giant cell** (right) **resulting from infection with respiratory syncytial virus. (d) Infection of a line of human lymphocytes with human immunodeficiency virus (HIV-1)**. Multinucleate giant cells are arrowed. (Parts **(a)** and **(c)** are reproduced with permission from Robinson, T.W.E. and Heath, R.B. (1983). *Virus diseases and the skin*. Churchill Livingstone, London.)

5   DETECTION OF ANTIVIRAL ANTIBODIES

It must be emphasized that isolation of a particular virus, although suggestive of the diagnosis, does not always prove a causal association between it and the patient's illness. As we saw in Chapter 4, some viruses are shed from clinically normal people. This uncertainty factor can be greatly diminished by the use of serological tests, in which a rising titre of antibody to a particular virus is sought, or serum is tested for the presence of specific IgM antibody. The first method depends on testing paired samples of serum, taken as soon as possible after onset and 10–14 days later; a fourfold or greater rise in titre is considered significant. The second, now widely used, has the advantage of rapidity in that specific IgM antibody is detectable a few days after the onset of illness (for an example, see Fig. 18.2).

5.1   *Class-specific (IgM) antibody tests*

We have already described some of the newer, rapid methods used for detecting viral antigens or antibodies. Among them, ELISA-type 'capture' methods (Fig. 27.2) are readily adaptable to the detection of specific antibodies, of which IgM is the most useful diagnostically. It is detectable within days of infection and remains so for 3–9 months, so that its finding is good evidence of a current or recent infection. In brief, the following steps are involved in testing for IgM antibody to a virus such as rubella (Fig. 27.2(b)).

1. IgG antibody to human IgM (anti-IgM) is adsorbed to a solid surface, e.g. a well in a microtitre plate.
2. The test serum is then added; IgM molecules are 'captured' by the anti-IgM.
3. Rubella antigen is added, and attaches only to rubella-specific IgM.
4, 5. Enzyme-labelled antibody to rubella is added and detected as described in Section 3.3.

Such tests are very reliable, provided that adequate controls are included and each step is followed by thorough washing to remove unbound, non-specific reagents. IgM antibody rises following secondary infections (e.g. reactivation of herpesviruses) or booster doses of polio or rubella vaccines are possible sources of error.

5.2   *Antibody analysis by Western blot*

The Southern blot method has no geographical connotation, but was named after the worker who invented a widely used method for DNA hybridization. A similar method used for RNA hybridization inevitably became known as Northern blotting, and Western blotting refers to its application for identifying

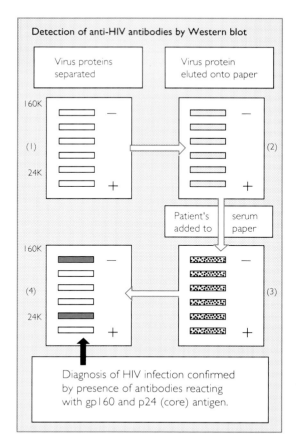

Fig. 27.5 Detection of anti-HIV antibodies by Western blot.

proteins. The development of such a technique for oligosaccharide antigens — surely to be known as Eastern blotting — is now a distinct possibility.

Since the correct diagnosis of HIV-1 infection has such important personal and social implications, it is necessary to confirm a positive result (see Chapter 22) by a different technique, usually Western blotting (or immunoblotting) (Fig. 27.5; numbers in the discussion which follows refer to this figure). Virus proteins are separated as bands according to their molecular weights by electrophoresis through a polyacrylamide gel (1). The bands are eluted ('blotted') on to chemically treated paper, to which they bind tightly (2). The test serum is added to the paper strip and any specific antibody attaches to the viral proteins (3). As in some other sandwich-type tests, an antihuman antibody labelled with an enzyme is added, followed by the enzyme substrate; the paper is then inspected for the presence of stained bands (4), which indicate the presence of complexes of specific antibody with antigen.

5.3 'Traditional' serological tests

These are still widely used and include complement fixation (many virus infections), radial haemolysis (screening test for rubella antibody), and occasionally haemagglutination-inhibition.

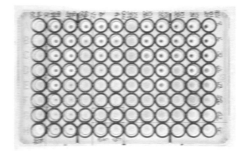

Fig. 27.6 Complement fixation tests.

Complement fixation test (CFT)

This reliable and versatile test has for many years served as the main diagnostic serological test. It is, however, relatively insensitive and requires large amounts of antigen, which are not available for all viruses.

Outline

1. The test serum is reacted with viral antigen and a defined amount of complement.
2. Specific antibody, if present, forms a complex with the antigen and complement.
3. Complement is then tested for by adding red blood cells sensitized with anti-red cell antibody. If it is still available, lysis of the red cells will take place (negative result).
4. If, however, the complement was previously mopped up by a specific viral antigen–antibody complex (see point 2), the red cells are not lysed by *their* antibody, and sink to the bottom of the well (positive result) (Fig. 27.6).

Application. Does not discriminate between IgM and IgG antibodies; used in hospital practice for routine serological tests on paired sera.

Radial haemolysis test

Outline

1. The virus is linked to sheep or human red blood cells by chromium chloride.
2. The treated cells are mixed with molten agarose which is poured into a petri dish or other suitable plate.
3. After cooling, small wells are punched in the agarose, each then being filled with a serum sample.
4. The plate is incubated overnight to allow diffusion of antibody into the agarose and combination with the antigen on the red cells.
5. A solution of complement is poured over the plate, and lyses those red cells on which both antigen and antibody are present.

Table 27.2 A typical year's work in a medium-size hospital diagnostic virology laboratory*

Virus	No. of samples examined	Diagnostic test
Hepatitis B	12 000	HBsAg
	300	HBe and anti-HBe
	2 000	anti HBs
	2 000	HBc IgG
	250	HBc IgM
Hepatitis A	1 000	HAV IgM
	100	HAV IgG
Rubella	5 000	IgG
	50	IgM
CMV	650	IgG
	750	IgM
EBV	1 100	EBNA-1 IgG/IgM
HIV	2 000	HIV Ab ELISA
	1 250	HIV p24 Ag ELISA
Many viruses	1 700	Cell culture
Herpes, diarrhoea viruses	600	Electron microscopy
Respiratory and herpes viruses	500	Immunofluorescence
Many viruses	500	Complement fixation

* The Royal London Hospital, with 450 beds and a catchment population of 250 000 persons.

6. Wells in which the test serum contained antibody are surrounded by clear zones of lysis, the diameter of which gives an indication of the amount of antibody present.

Application. This variant of the CFT can be used to test for influenza antibody, but is much more widely employed as a test for rubella antibodies in females. It is *not* accurate enough for assessing the antibody status of individual patients, but can be used to screen large numbers of sera, e.g. from antenatal clinics.

In conclusion, Table 27.2 gives an idea of the annual workload of the virology laboratory in a busy teaching hospital.

6   REMINDERS

◆   Modern, rapid methods of diagnosing virus infections are now making a significant contribution to the management of patients.

◆ The work of the laboratory is greatly helped by the provision of appropriate specimens and clinical information.

◆ **Rapid tests for viruses or their antigens** (result in 24 h) include **immunofluorescence**, **ELISA**, **latex agglutination**, and **electron microscopy**. The specificity of many of these tests is enhanced by the use of **monoclonal antibodies**.

◆ **Nucleic acid hybridization** and the **polymerase chain reaction** are highly sensitive methods for detecting viral genome in tissues, but their use is so far confined to specialist laboratories.

◆ **Virus isolation** in cell cultures is still useful on occasion, but is slow and becoming superseded by rapid methods.

◆ **Detection of viral antibody** provides a good element of specificity in the diagnosis. Demonstration of a significant rise in titre is necessarily a slow method, and is being overtaken by tests for **specific IgM antibody**.

CHAPTER TWENTY-EIGHT

CONTROL OF VIRAL DISEASES BY IMMUNIZATION

1   INTRODUCTION

1.1   *Historical*

Not until the germ theory of disease was accepted in the late nineteenth century and concepts of immunity and immunochemistry were developed did it become scientifically conceivable that infection by viruses could be prevented by injection of killed (inactivated) viruses. These immunizing preparations were called vaccines. They contain what in modern terminology are virus proteins with antigenic areas called **epitopes**. Viral epitopes induce B cell and T cell immunity. Viruses with reduced or attenuated virulence in 'live' vaccines replicate at the site of injection to produce these viral proteins. Hundreds of years before the new era of European scientific discovery, clinical observation in the Middle East and China had led to an empirical concept of **variolation**, whereby inoculation of the infective material from pocks of smallpox sufferers was used to prevent attacks of the disease itself. The subsequent technique of 'vaccination' (Latin *vacca*: cow) was a significant development of this basic idea: milkmaids and farmers who had been infected previously with cowpox were noted to be subsequently immune to attacks of smallpox. We realize now that the explanation for this protection lies in the antigenic similarities between these two poxviruses. A mutant of poxvirus, called vaccinia, was used in modern times to eradicate smallpox. The word 'vaccine' has come into general use to describe all these immunizing preparations, whether live or inactivated.

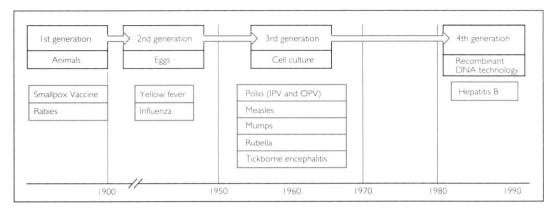

Fig. 28.1 Four generations of virus vaccines.

Table 28.1 Virus vaccines currently licensed in the UK

Vaccine	Source of virus	Inactivated or attenuated vaccine	Route of administration	Comments
Vaccinia	Lymph from scarified animal skin	Attenuated	Scarification of arm	Stocked for emergency use only; US and CIS (USSR) armies immunized
Yellow fever	Eggs	Attenuated	i.m.	Long immunity induced
Influenza	Eggs	Inactivated (subunit)	i.m.	Antigenic variation of virus outdates vaccine yearly; 70 per cent effective
Polio	Monkey kidney or human diploid cells	Inactivated Attenuated	i.m. Oral	Both vaccines are highly efficacious
Measles	Chick embryo cells	Attenuated	i.m.	Successful triple vaccine (MMR)
Rubella	Human diploid cells	Attenuated	i.m.	with at least 90 per cent efficacy;
Mumps	Chick embryo cells	Attenuated	i.m	long-lasting immunity
Rabies	Animal brain	Inactivated	i.m.	Postexposure prophylaxis only
	Human diploid cells	Inactivated	i.m.	Pre- and postexposure prophylaxis
Hepatitis B	Human serum	Inactivated (subunit)	i.m.	
	Yeast (recombinant)	Subunit	i.m.	Recombinant DNA vaccine

i.m., intramuscular.

The first truly scientific vaccine, against rabies, was pioneered by Pasteur in 1884; and other milestones of vaccine development include attenuated yellow fever virus vaccine developed by the French and Americans in the 1930s (Fig. 28.1).

Now, in the early 1990s, we have reached a new era and the fourth generation of immunizing agents, those produced with recombinant DNA technology, e.g. hepatitis B subunit vaccine. Some of the features of these vaccines are listed in Table 28.1.

1.2 The WHO programme: 'Health for all by the year 2000'

As a result of mass immunization campaigns during the last two decades, childhood infections such as polio, measles, and rubella are well controlled in many of the more wealthy countries of the world. But it should also be remembered that in a global context infectious diseases still take a heavy toll both in mortality and general human suffering. Paralytic polio is still widespread in South America, Asia, and Africa, and measles causes very serious problems in children in these countries. Nevertheless, the scientific problems faced in development and production of safe and efficacious vaccines for these diseases have been overcome; only the political will and economic resources to use them in developing countries are now needed. As long as reservoirs of any of the viruses remain, importations into apparently 'virus-free' countries will be a constant health problem. The World Health Organization (WHO) has now included polio and measles eradication in the international campaign 'Health for all by the year 2000'.

Smallpox eradication is the best example of a vigorous international approach to the elimination of an important infectious disease. It was initiated by the USSR (now CIS) at a WHO meeting in Alma Ata in the late 1960s and supported fully by the USA; with the necessary funds and scientific expertise the world vaccination campaign was pushed forward by WHO in some of the poorest nations on earth with dramatic success. Similar international co-operation could now result in the vanquishing of polio and measles.

Preventive immunization is extremely cost-effective in savings from hospitalization and medical treatment. This is particularly the case with polio, measles, and rubella, which sometimes give rise to long-term deleterious effects.

2 THE PRACTICALITIES AND TECHNOLOGY OF VIRUS VACCINE DEVELOPMENT

2.1 Virus vaccine production and standardization

Vigorous efforts are made by international organizations such as WHO to ensure that virus vaccine **seed material** and production facilities in different countries conform to given standards. Thus, most vaccine manufacturers use well characterized stock vaccine viruses and cell cultures for preparing vaccines, and proper quality control of the final products. A **seed virus technique** is used whereby a large batch of vaccine virus is frozen at -70°C. From this, the actual vaccine is produced by infecting cells with virus that is only one or two passages removed from the seed. This procedure reduces the opportunity for further random mutations to occur in vaccine virus, which might otherwise alter its virulence or antigenic characteristics.

The quantity of infective virus in a live vaccine is quantified in cell cultures to ensure correct **infectivity titre** and the final batch of vaccine is tested rigorously to exclude the presence of viral or bacterial contamination.

2.2   *Choice of cell substrates*

John Enders and his colleagues achieved a major scientific breakthrough in the 1940s when they reported that a strain of polio virus could be cultivated in monkey kidney cells which, unlike neural tissue, can readily be grown in the laboratory. The immediate result of this observation (together with the discovery of antibiotics, which prevented bacterial contamination of cell cultures) was the development of effective vaccines against polio. In general, the criteria for selection of cells for vaccines are ready **availability**, the **lack of potential oncogenicity**, **genetic stability**, and **freedom from** demonstrable contamination with **extraneous viruses**. For most of the earlier vaccines, such as polio and rubella, cells from a variety of mammals were used but, with the growing appreciation of the presence of retroviruses in animal cells and perhaps other, as yet undetected, viruses, together with growing concern about the use of primates, more vaccine viruses are now cultivated in human cells. These are **diploid**, with a limited lifespan in culture, as in nature (Chapter 3). The cells can be passaged only 40 to 50 times *in vitro* before dying out so that, in practice, large batches are frozen in liquid nitrogen at an early passage and are available for manufacturers to produce virus vaccines.

2.3   *Problems and contra-indications*

Although the development of effective virus vaccines is one of the major successes of biomedical research, some serious problems have come to light and there have been some bitter failures.

As an example of a totally unexpected problem, an inactivated vaccine against respiratory syncytial virus resulted in some immunized children developing a more serious infection than their non-immunized classmates, when they were later in contact with virulent virus (see Chapter 9). Another problem encountered with some attenuated vaccines such as those for polio and influenza is **reversion** to parental-type virulence. Such events provide a warning that we still do not fully understand the underlying genetic mechanisms determining virulence or attenuation of most viruses.

No biological preparation administered to humans can be completely free of unwanted side-effects, and vaccines are no exception. But, compared to the natural diseases against which these vaccines provide very effective protection, the side-effects are minimal, as can be seen in Table 28.2. It is,

Table 28.2 Main side-effects and contra-indications for licensed viral vaccines

Vaccine	Potential side-effects	Main contra-indications for vaccination
Inactivated vaccines		
Influenza	Local reactions including redness at inoculation site. Guillain–Barré syndrome is exceedingly rare	Serious egg allergy
Rabies	Mild local reactions with modern human diploid cell vaccine. Local and systemic reactions including encephalomyelitis with older animal brain vaccines	None
Polio	None	None
Hepatitis B	None	None
Live attenuated vaccines		
Measles	Mild. Malaise, rash, fever, headache in a low proportion	
Rubella	Mild. Lymphadenopathy and joint pain in a low proportion	Pregnancy; the immunocompromised; serious egg allergy (except rubella and polio)
Mumps	Mild. Fever and parotitis. Post-vaccination meningitis with Urabe strain	
Yellow fever	Mild. Malaise, headache in a low proportion	
Polio	Vaccine-associated paralysis; exceedingly rare	

however, important to remember that live attenuated vaccines should not be given to pregnant women for fear of infecting the fetus or to immunocompromised persons in whom they might cause actual disease.

3 | ADMINISTRATION OF VIRUS VACCINES

Apart from the immunization strategy of the international WHO programme 'Health for all by the year 2000', most national health authorities organize childhood immunization schedules. The details vary from country to country; the UK schedule is summarized in Table 28.3; it now incorporates the triple mumps, measles, and rubella vaccines (MMR). Polio, of course, is an important vaccine and the attenuated version is used in the UK. Some Scandinavian countries use the inactivated polio vaccine. It is most important that these national campaigns are conducted in a vigorous manner to achieve immunization rates greater than 90 per cent in childhood. Sometimes it is impossible to predict at the outset whether a vaccination campaign will produce enough

Table 28.3 Schedules for immunization with virus vaccines in the UK

Vaccine	Age of vaccinee	Comment
Polio – 1st dose	2 months	Live attenuated
– 2nd dose	3 months	Live attenuated
– 3rd dose	4 months	Live attenuated
– 4th dose	4·5 years	Live attenuated
– 5th dose	15·18 years	Live attenuated
Measles, mumps and rubella (MMR)	13·15 months	Live attenuated
Rubella	Adult	Booster dose for 10·14 year old girls. For non-immune women before pregnancy or after delivery.
Influenza A and B	Adult	High risk group (Chapter 10)
Hepatitis B*	Adult	High risk group (Chapter 20)
Polio	Adult	Booster dose for 15·18 year olds and for those travelling to endemic areas.
Yellow fever	Adult	Before visits to endemic areas. (Valid certification of immunization is required)
Rabies	Adult	Persons in high risk occupation, (e.g. quarantine unit) may be given vaccine prophylactically in two doses.

herd immunity to prevent spread of the natural virus. Figure 18.3 shows the results of two successive strategies to prevent rubella epidemics, by immunizing either adolescent girls or babies of both sexes. The latter approach is by far the more successful. The remaining vaccines licensed in the UK are administered to persons at special risk of contracting a particular viral disease, such as elderly persons with influenza, health care staff with hepatitis B, and veterinarians with rabies vaccines.

Many queries are received in virus units in the UK each year about immunization requirements for foreign travel. Polio vaccine is often forgotten by travellers and yet vaccine-induced immunity does decline and the adult traveller may be at risk of contracting the infection in many areas of the Middle East, Asia, South and Central America, and Africa. Yellow fever may be contracted during visits to endemic areas and immunization is obligatory for entry to these regions and subsequent travel to uninfected countries. Prophylactic rabies vaccine is not recommended unless the traveller is visiting an endemic area for an extended period.

Virus vaccines must be kept cool in a refrigerator. Live vaccines are particularly susceptible to heat-inactivation and are thus often transported as a freeze-dried powder to be reconstituted with sterile water at the time of use. Live polio vaccines are administered as oral drops, usually on a sugar lump, and all the other current viral vaccines, whether live or inactivated, are administered by intramuscular injection.

4 PASSIVE IMMUNIZATION

Injection of human immunoglobulin preparations containing various antibodies give immediate partial or complete protection against infection by certain viruses. Table 28.4 summarizes the preparations currently available in the UK. This is by no means a universal protective method because administration of antiviral antibodies to persons already infected with certain viruses could actually make the infection worse: with dengue, for example, antibodies may form complexes and provoke an untoward reaction (Chapter 17).

Table 28.4 Human immunoglobulins used for passive prophylaxis

Preparation	Comments
Normal human immunoglobulin (HIG)	For prevention or modification of measles in persons at special risk after contact with an infection, e.g. immunocompromised children or adults For prevention or modification of hepatitis A in travellers to endemic countries excluding Europe, USA, and Australasia
Hepatitis B immunoglobulin (HBIG)	Can be co-administered with vaccine to provide rapid protection. Administered to persons with needlestick injuries. Not available for travellers
Human rabies immunoglobulin (HRIG)	To provide rapid protection after exposure to virus until vaccine immunity develops
Zoster immunoglobulin (ZIG)	For immunosuppressed or leukaemia patients; neonates or pregnant contacts of cases
Lassa convalescent plasma	Used therapeutically

5 VIRUS INFECTIONS YET TO BE CONTROLLED BY IMMUNIZATION AND THE NEW TECHNIQUES OF MOLECULAR BIOLOGY

Virologists are now left with the difficult 'problem' viruses to control by vaccines: influenza, common cold viruses, respiratory syncytial viruses, arboviruses, herpesviruses, and HIV. The immunogenic proteins of influenza undergo antigenic drift (see Chapter 10), so the virus can circumvent both vaccine-induced and natural immunity. We would expect similar problems with an AIDS vaccine. A formidable task with HIV is to prevent integration of the proviral genome into susceptible cells. There are over 100 serotypes of rhinovirus, far more than could be accommodated in a common-cold vaccine.

5.1 *Genetic engineering*

New ideas and techniques of molecular biology have been introduced recently into vaccine technology. Thus, as with the new hepatitis B vaccine, viral genes or portions thereof can now be cloned in plasmid vectors and transferred to bacteria, yeast, plant, or human cells. In the example of cloning illustrated in Fig. 3.4, a viral gene coding for an immunogenic protein is excised from the virus and cloned into a plasmid vector, which is then inserted into a bacterial cell. The plasmid replicates alongside the bacterium and provides the necessary genetic information for synthesis of the viral protein. This protein may be produced in large quantities in a bacterial fermentation apparatus and purified from the bacterial broth. Alternatively, the vector may be cloned into eukaryotic cells or into insect cells or larvae for efficient expression of the viral protein.

Perhaps the most exciting molecular technique is the cloning of viral genetic information into the genome of other large DNA viruses, such as vaccinia or adenovirus. Genes of immunogenic proteins of rabies, influenza, HIV, and hepatitis have been cloned into the thymidine kinase (*TK*) locus of the genome of vaccinia virus. Insertion of new genetic information at this point in the vaccinia genome is expected to attenuate this virus further. The vaccinia virus genome is so large that its functions are not compromised by excision of a portion and reintroduction of a new section of nucleotides. After inoculation into the human skin, vaccinia virus replicates and concomitantly produces the new proteins of the cloned viral gene. The vaccinated person develops immunity to vaccinia and to the cloned viral protein. The first experiments have been performed with the HIV-1 gp160 gene and, after vaccination, the volunteers developed both T cell and B cell immunity to the HIV protein. There are, however, worries that renewed large-scale use of vaccinia, even with cloned genes in the *TK* locus, could result in serious side-effects in vaccinees, including postvaccination encephalomyelopathy. The use of adenovirus as a vector is being explored as an attractive alternative to vaccinia. It has a large DNA genome, and can be administered orally with no known side-effects.

5.2 *Short peptides as vaccines*

Our understanding of the nature of the epitopes (antigenic determinants) for B and T cells on virus proteins is progressing. As virus proteins are defined by the techniques of molecular biology the expectation is that normally quiescent viral epitopes which possess cross-reactive antigenic properties will be uncovered and exploited as novel immunogens. Some viral epitopes consist of as few as eight amino acids. Cattle can be protected against foot and mouth disease virus by immunization with short synthetic peptides of the correct amino-acid sequence. In theory, many such epitopes for different viruses could be synthesized and linked together as a single immunogenic protein able to induce immunity to a wide range of infective agents.

5.3 *Adjuvants*

Adjuvants are molecules which can prolong and enhance the immune response to inactivated or subunit vaccines. Aluminium hydroxide is one of the few adjuvants currently licensed for use in humans. Methods of presentation of viral antigens to the immune cells are important and immunogenicity can be enhanced if viral proteins are aggregated in novel ways with or without adjuvants. Aggregation of viral proteins by saponin molecules results in formation of immune-stimulating complexes (ISCOMS); and viral proteins have been incorporated into lipid spheres (liposomes) containing muramyl dipeptides which increase the immunological response and memory of the host.

6 REMINDERS

◆ Viral vaccines prevent infection by antigenic stimulation of the host, resulting in the generation of **neutralizing antibody** and **cytotoxic T cells**. Immunity starts to develop some days after vaccination.

◆ The vaccine may be a chemically inactivated suspension of **whole or fractionated virions** containing immunogenic proteins, or a **live attenuated** preparation which replicates to produce the necessary immunogens.

◆ There are effective live attenuated vaccines against measles, mumps, rubella, polio, and yellow fever and inactivated vaccines against influenza, rabies, polio, and hepatitis. These vaccines are very cost-effective.

◆ Vaccines are not without side-effects but these are extremely mild compared to the natural disease. **Pregnancy and immunocompromised states are contra-indications to live attenuated vaccines**.

◆ Problems encountered during vaccine development have been reversion to virulence of attenuated polio viruses, failure to ensure complete chemical inactivation of polio viral infectivity, antigenic changes in the field virus, particularly with influenza, and enhanced disease in children immunized with inactivated measles and respiratory syncytial virus vaccines.

◆ The first example of the use of **recombinant DNA technology** in vaccine production is **hepatitis B antigen** cloned in yeast cells. Viral vectors, such as vaccinia or adenovirus, are under consideration as new delivery vehicles for multiple virus antigens.

◆ Certain viruses remain uncontrolled by immunization because of **multiplicity of serotypes**, i.e. rhinoviruses and arboviruses, or **major antigenic change**, i.e. in influenza and HIV. Furthermore, a fully effective vaccine against HIV would have to prevent DNA integration into the host-cell chromosomes.

◆ Passive prophylaxis with **human immunoglobulin preparations** gives a measure of immediate protection against infection by measles, hepatitis A and B, rabies, VZV, and Lassa. Immunoglobulin may be co-administered with inactivated vaccines such as rabies and hepatitis B.

CHAPTER TWENTY-NINE

ANTIVIRAL CHEMOTHERAPY

1 INTRODUCTION

Certain important virus diseases such as polio, measles, and rubella can be kept under good control with live virus vaccines. But, as we saw in the previous chapter, it would be difficult to imagine the development of successful vaccines for many viruses, because of a multiplicity of serotypes, or variability or complexity of their antigenic structure. These are the target viruses for antiviral chemotherapy. There are currently 10 antiviral compounds licensed for use in the UK against herpes, HIV-1, respiratory syncytial virus (RSV), and influenza A virus. A negative feature of all the known antivirals is their very narrow spectra of activity; thus antiherpes compounds have no effect against influenza. All the drugs have been discovered by random biological screening in the laboratory.

But would it not be possible to devise a universal antiviral against all viruses or does such a compound exist already as interferon? The answer is unfortunately no. The great hopes raised over 30 years ago when interferon was discovered by Alick Isaacs and Jean Lindenmann have not been realized, and its applications are still very limited.

2 POINTS OF ACTION OF ANTIVIRALS IN THE VIRUS LIFE CYCLE

Scientists have searched for the last 30 years for molecules which would inhibit virus-directed events rather than normal cellular activities. This task is not easy. The potential points of inhibitory action of antiviral drugs are:

◆ **Binding to the free virus particle**. This represents a rather unusual target but could stop virus infection at a very early stage.

◆ **Interference with virus adsorption or attachment to the receptor binding site on the cell**. Synthetic cell receptors could act as 'decoys' and hence prevent infection of cells. An example is the possible use of soluble CD4 to prevent HIV infection.

◆ **Inhibition of virus uncoating and release of nucleic acid**.

◆ **Inhibition of viral nucleic acid transcription and replication**. Certain viruses code for specific enzymes of their own, such as influenza (RNA transcriptase), herpes (thymidine kinase and DNA polymerase),

Table 29.1 Examples of antiviral drugs affecting different steps in virus multiplication

Target	Drug	Virus inhibited
Viral adsorption	Soluble CD4	HIV-1
Penetration	Amantadine*	Influenza A
Uncoating	Arildone	Picornavirus, herpes simplex
Viral nucleic acid synthesis	Acyclovir*	Herpes simplex and varicella zoster
	Adenosine arabinoside*	Herpes simplex and varicella zoster
	Ribavirin*	Influenza A and B Togaviruses
	Trifluorothymidine*	Herpes simplex
	Azidothymidine*	HIV-1
	Dideoxyinosine	HIV-1
	Interferon*	Range of viruses
	Foscarnet*	HIV-1 and herpes simplex
Binding to intact virus particle	Disoxaril	Rhinoviruses
Virus release	Amantadine*	Influenza A

* Licensed antiviral.

and HIV (reverse transcriptase, integrase, and protease). These viral enzymes form important targets for inhibition.

◆ **Interference with cellular processing of viral polypeptides**, by preventing addition of sugar or acyl groups. This may be the least successful approach because cellular and viral proteins are processed in similar ways.

◆ **Prevention of virus budding**.

Some of the current antivirals (Table 29.1) are very effective inhibitors and, although few in number compared to antibacterials, are now being used widely in hospitals and in general practice. The chemical structures of these drugs are illustrated in Fig. 29.1.

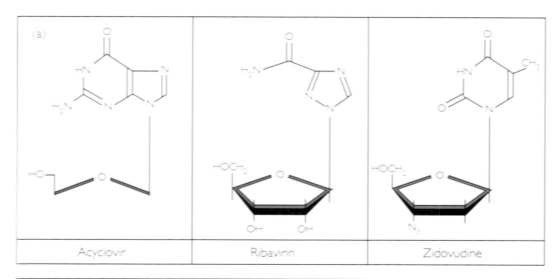

| Acyclovir | Ribavirin | Zidovudine |

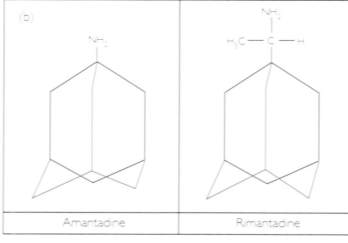

| Amantadine | Rimantadine |

Fig. 29.1 The chemical structures of some antiviral compounds. **(a)** Acyclovir (HSV, VZV), ribavirin (RSV), azidothymidine (HIV-1); **(b)** Amantadine (influenza A), rimantadine (influenza A).

3 THE CLINICAL USE OF ANTIVIRALS

A listing of antivirals according to their clinical usefulness and effectiveness would be headed by acyclovir against herpes simplex viruses type 1 and 2, followed by amantadine against influenza A virus, zidovudine against HIV-1, and ribavirin against RSV infection in children (Table 29.2)

Table 29.2 Examples of treatment of virus infections with antivirals

Infection	Drug	Dosage suggestions
Herpes encephalitis	Acyclovir	10 mg/kg* 8-hourly by intravenous infusion
Serious generalized HSV in immunocompromised	Acyclovir	10 mg/kg 8-hourly by intravenous infusion for 1 week or 200 mg orally five times daily
Prevention of recurrent HSV	Acyclovir	Oral drug 400–800 mg daily
Keratoconjunctivitis	Acyclovir	3% eye ointment
First episode of genital HSV	Acyclovir	Oral drug 200 mg five times daily for 5–10 days
Severe VZV	Acyclovir	10 mg/kg 8-hourly by intravenous infusion for 1 week or 800 mg five times per day by oral route
Influenza A	Amantadine	Oral drug 100 mg daily for 10 days prophylactically or therapeutically. The prophylactic use may continue for 5 weeks
AIDS	Zidovudine**	Oral drug 1200 mg per day
Asymptomatic HIV-1	Zidovudine	Oral drug 500 mg per day
Severe CMV in immunocompromised patients	Ganciclovir	5 mg/kg 12-hourly by intravenous infusion
Severe RSV infections in children	Ribavirin	20 mg/ml solution administered with small-particle aerosol generator 12-18 hourly for 3–7 days

* Body weight
** Combination therapy with didanosine (ddI) is currently practised.

3.1 *Therapy and prophylaxis*

Antivirals are mostly used **therapeutically** and are administered as soon as possible or after the first clinical signs of infection are noted.

To a lesser extent antivirals are used to prevent virus infections, in much the same way as most vaccines are used. But **prophylaxis** with a chemical antiviral has the advantage of speed of action, since some protection would be anticipated within 1 h of drug administration. It should be remembered that the converse is also correct and that immediately drug prophylaxis is discontinued the patient becomes susceptible to virus infection.

Following oral or intravenous administration, effective tissue concentrations of the drug may be achieved in minutes but often the half-life of the drug is only a few hours and part is often excreted unchanged in the urine.

3.2   Side-effects

All drugs have side-effects and antiviral compounds are no exception. Even the very safe antiherpes drug acyclovir can cause gastrointestinal symptoms such as nausea and vomiting; particular care must be exercised in patients with renal failure since, even in normal circumstances, about 70 per cent of the drug is excreted unchanged in the urine. Severe renal malfunction can result in undesirably high concentrations of drug in the blood. Ganciclovir induces more serious side-effects and neutropenia occurs in one-third of patients; thrombocytopenia, rash, and nausea are also important side-effects. Zidovudine in high dosage induces anaemia, neutropenia, and leukopenia in approximately one-third of patients. In a few patients amantadine causes the slight neurological effect of 'jitteriness' which ceases immediately the drug is discontinued. Even interferons are not free of side-effects such as psychiatric changes, fatigue, depression, and even severe somnolence.

An important contra-indication for these drugs is pregnancy, because of potential teratogenic effects.

3.3   Acyclovir against herpesviruses

Clinical application

The nucleoside analogue predecessors of acyclovir such as idoxuridine (IDU), and trifluorothymidine (TFT) were useful for treatment of herpetic eye infections, particularly keratitis, and adenine arabinoside (ara A) had a role in the treatment of herpes simplex encephalitis and serious paediatric infections. But acyclovir is so much more effective that its use has transformed the often difficult clinical management of serious herpetic infections. Acyclovir is used in the prophylaxis of herpes simplex and zoster infections in bone marrow and heart transplant patients, and therapeutically to prevent their spread in those already infected. It is also very effective in herpes simplex encephalitis, if administered early in the disease. Continuous treatment with acyclovir prevents recurrent HSV-1 and 2 infections, particularly those of the genital tract. Higher doses of the drug are used to treat severe varicella-zoster in the elderly

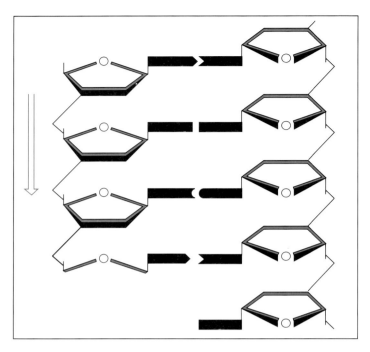

Fig. 29.2 DNA chain termination by acyclovir. Acyclovir (ACV) lacks a 3' OH group; when it is incorporated into a DNA chain, further elongation of the chain is prevented because no 3' OH group is present to form a phosphodiester linkage with the next nucleotide.

and in immunocompromised patients (Table 29.2). A derivative, ganciclovir, has some antiviral activity against cytomegalovirus (CMV) and is used to treat life-threatening pneumonia after bone marrow transplants. A completely unrelated compound, foscarnet, is used to treat CMV retinitis.

Mode of action

Acyclovir possesses a combination of biochemical and pharmaceutical properties which explains its unique antiherpes virus specificity. First, the compound is **phosphorylated to the monophosphate only in herpes-infected cells**, since this step requires a **herpes thymidine kinase (TK)** and cannot be achieved by normal cellular TK. The viral TK is less 'precise' than the corresponding cellular TK and so, unlike the latter, will accept 'fraudulent' substrates such as acyclovir. **Phosphorylation to the di- and triphosphate is achieved by cellular enzymes**. The second specific feature of the drug is that the **triphosphate**, the active moiety, binds to and specifically **inhibits the function of the herpesvirus DNA polymerase** (Fig. 29.2). It has little effect on normal cell DNA polymerase and hence is not toxic. However, a latently infected cell cannot be cleared of virus, so that acyclovir does not eradicate herpes infection, although it can prevent clinical recurrences. Acyclovir also has DNA chain termination activity and so might, at least in theory, inhibit DNA replication in uninfected cells, but, if this happens at all, the effect must be very slight and the compound is considered to be safe in clinical use: patients with recurrent herpes lesions have been effectively treated with daily doses for several years without side-effects.

3.4 Amantadine and rimantadine against influenza A virus

Clinical application

■■■■■ Prophylactic administration will prevent illness caused by influenza A viruses in 80 per cent of individuals. Unfortunately, it has no effect against influenza B viruses and so cannot be used as a substitute for influenza vaccine in the 'special risk' group (see Chapter 10). A very similar molecule but with an extra methyl group (rimantadine) (Fig. 29.1) has equivalent clinical activity against the virus but causes fewer side-effects.

Studies in prisons, schools, and universities showed that, if amantadine is given within 24 h of onset, influenza resolves more quickly than in untreated people and the period of incapacity is reduced. Current recommendations from the World Health Organization are that the compound may be used prophylactically when the presence of influenza A virus in the community is confirmed. **Prophylactic use** can continue for 5 weeks or until the end of the epidemic is in sight. As with influenza vaccines, chemoprophylaxis is recommended only for the '**special-risk groups**', such as the over-65s, diabetics, and persons with chronic heart or chest diseases who have not been immunized or who wish to receive extra protection. The compound can be used **therapeutically** in the same special-risk groups.

Mode of action

■■■■■ The antiviral action of amantadine is mediated by its ability to **increase the pH of intracellular vacuoles**. Influenza A virus normally infects cells by catalysing the fusion of its viral membrane with a cellular membrane in intracellular vacuoles at **low pH** (Chapter 10). If the vacuolar pH is raised by amantadine, virus-induced fusion is prevented and subsequent release of viral nucleic acid and hence viral infection is blocked. One of the structural proteins of virus, the M2 protein, normally acts as an ion channel, allowing passage of hydrogen ions to the interior of the virus. Amantadine binds to this protein and blocks the channel, much as a gate blocks a passageway, and acidification cannot occur.

3.5 Zidovudine against HIV-1

Clinical application

■■■■■ Within 2 years of the isolation of HIV-1 a series of antiviral compounds had been discovered and one of them, zidovudine (azidothymidine), was shown to be effective in prolonging the lives of AIDS patients by a year or so. A double-blind placebo-controlled trial of zidovudine in AIDS and ARC patients had to be stopped and the code broken when it was found that mortality in the zidovudine-treated group was 1, compared to 19 in the control group, out of a combined total of 281 patients. However, this beneficial effect is not permanent. In addition, fewer opportunistic infections developed in the zidovudine-treated group and a reduction in the degree of p24 antigenaemia indicated a specific antiviral effect of the compound. However, high dose zidovudine has marked toxic effects on bone marrow cells that necessitate transfusion in one-third of patients. Some HIV-1 strains have mutated to become drug-resistant. Thus the search has intensified, both for entirely new drugs and for other dideoxy nucleoside analogues that act synergistically with zidovudine, so that the dosage may be lowered to avoid side-effects.

Mode of action

As with acyclovir, the molecule has to be **phosphorylated** intracellularly to produce the active antiviral drug. The triphosphate is a very potent inhibitor of viral RT enzyme and prevents nucleotide chain elongation. The 3' positioning of the azido group blocks the essential phosphodiester linkage which would normally enable the next nucleotide to be added to the growing DNA chain. Azidothymidine triphosphate binds to the **viral reverse transcriptase** rather than to the cellular DNA polymerase, giving some specificity of action. By contrast with acyclovir, cell enzymes phosphorylate the molecule and hence intracellular concentrations of active drug increase in normal cells; this partly explains its toxic effects.

3.6   Interferon

Clinical application

The use of interferon in the clinic was for long restricted by the small quantities available. With the development of cDNA cloning technology the situation changed and controlled trials of genetically engineered interferons have now been conducted. Three interferon preparations are licensed, alpha 2b, alpha 2a, and alpha N1, all for treatment of persons with hairy cell leukaemia (Chapter 22). The limited knowledge of concentrations of interferon in tissues during natural viral infections has impeded investigations of the therapeutic efficacy of exogenous interferon. Successful trials of recombinant interferons have been carried out in volunteers infected with rhinoviruses and in patients with hepatitis B. But the original hopes that interferon would prove to be a universal antiviral drug have not been fulfilled. Interferon may have a future as an immune modulator. Isoprinosine has been licensed for this purpose but evidence of a significant clinical effect is poor. Compounds which could increase or even restore immune function are urgently required, particularly for AIDS patients.

Mode of action

There are three main types of human interferons called alpha, beta, and gamma and there are several different species within alpha and beta types of interferons. The three types of interferon are produced in leukocytes, fibroblasts, and immune T cells, respectively.

Interferon was first detected by its antiviral action in cell cultures infected with influenza virus, but it is now known that its synthesis can be induced by a number of agents in addition to viruses, e.g. bacteria, dsRNA, and mitogens. The precise mechanism by which cells are made refractory to infection by viruses has not yet been established with certainty, but Fig. 5.5 summarizes the main features.

4 THE FUTURE

Although chance will continue to play a role in the discovery of new antivirals, knowledge of viral gene structure and functioning could now lead to a new generation of drugs. To give an example, the interaction between an antiviral compound (disoxaril) and a virus protein has been visualized at the atomic level by X-ray crystallography. A small molecular-weight inhibitor of the common cold virus has been identified in a 'cave' at the bottom of the receptor-binding pocket of the virus. This drug–virus interaction blocks a pore in the virion through which ions would normally pass to its interior to aid uncoating and release of viral RNA. In the presence of the drug, virus uncoating and infection of cells are aborted. Such computer-aided studies at the atomic level could allow the **design** of new inhibitors. In parallel with the search for new drugs, more intensive study of existing ones could give clues to the synthesis of molecular varieties with **broader spectra of antiviral activity**.

The emergence of **drug-resistant viruses**, a major consideration in antibacterial chemotherapy, will have to be closely monitored. Strains of HIV resistant to zidovudine have been recovered from AIDS patients who have been treated for 6 months or longer with the drug. Herpesviruses resistant to acyclovir have been isolated and may be a problem in immunosuppressed patients. These drug-resistant herpes variants either have no TK enzyme or have a genetic change in the viral *TK* gene. The altered TK does not phosphorylate acyclovir and so the antiviral triphosphate molecule is not synthesized in the virus-infected cell. After a few days of treatment with amantadine, drug-resistant strains of influenza A with specific nucleotide changes in the *M2* gene may be detectable. There is, however, no documented evidence of spread in the community of a drug-resistant virus: such viruses may even have reduced virulence.

The future of antiviral chemotherapy holds promise of more rapid development of new molecules, major advances in targeting to viral genes and virus-infected cells, and even excision of integrated viral genetic material.

5 REMINDERS

◆ Vulnerable points for attack in the virus life cycle are adsorption to cells, penetration, **replication of viral genome**, and budding. Viral enzymes such as RNA transcriptase and DNA polymerase, protease, and reverse transcriptase are excellent targets.

◆ Antiviral compounds are usually used therapeutically but prophylactic use is sometimes possible. Unlike immunization, administration of an antiviral can give rapid protection against infection whereas its premature discontinuation may allow the infection to recur.

◆ Ten antiviral drugs are licensed for clinical use: acyclovir, azidothymidine, amantadine, ganciclovir, ribavirin, idoxuridine, adenine arabinoside, trifluorothymidine, foscarnet, and interferon.

◆ Acyclovir and zidovudine are pro-drugs, the active form being the nucleoside **triphosphate**. Acyclovir requires a **herpes-coded TK enzyme** to add the first phosphate and hence the antiviral molecule only accumulates in a virus-infected cell. The triphosphate specifically inhibits the **herpes DNA polymerase**. Zidovudine is phosphorylated in both normal and virus-infected cells and so is less specific in inhibitory activity. Both molecules have some DNA chain termination activity.

◆ **Drug-resistant influenza**, **herpes simplex**, and **AIDS** viruses have been isolated and pose potential problems, particularly in immunodeficient patients.

SAFETY PRECAUTIONS: CODES OF PRACTICE, DISINFECTION, AND STERILIZATION

1   SAFETY IN THE LABORATORY

There have been many instances of laboratory workers — some famous microbiologists among them — infecting themselves with serious or even fatal consequences. Such incidents are now comparatively rare, but only because of greater awareness of the dangers, and the rigorous implementation of safety precautions.

In the UK, microbiological and other laboratories must conform with the relevant provisions of the Health and Safety at Work Act and other statutory requirements, which are supplemented by local rules that apply to the circumstances of individual laboratories. These codes of practice must be readily available to all staff coming into contact with infectious or potentially infectious patients or specimens. Since they are quite lengthy and in any event vary somewhat from place to place, they cannot readily be summarized; at the end of this appendix we have listed some publications that provide detailed guidance. There is however, one golden rule that must be mentioned.

◆ **_All_ microbiology specimens — including blood samples — must be regarded as potentially infective and handled accordingly.**

2 SAFETY IN MEDICAL AND SURGICAL DEPARTMENTS

Every hospital should have a Control of Infection Officer, who is often a microbiologist, and a written Control of Infection Policy covering such matters as isolation of infectious and immunocompromised patients, disinfection methods, and procedures for dealing with hazardous microbial diseases including those due to viruses—AIDS, hepatitis B, and haemorrhagic fevers. There should also be contingency plans for dealing with outbreaks of infection within the hospital. This information must be available to all staff coming into contact with patients.

3 SAFETY IN DENTAL DEPARTMENTS

Microbiological safety within dental units presents special problems: students are advised to read the appropriate chapters in the book by Macfarlane and Samaranayake, details of which are given in Appendix C. It provides a comprehensive and practical account of this topic, including the special precautions to be taken when dealing with high-risk patients. Here we shall mention only the following general points.

3.1 *Infection of dental staff by patients*

It is usually possible to identify potentially infectious medical patients, but this is not so with dental patients, any of whom is liable to be shedding one or another virus from the mouth, often without symptoms. Viral infections associated with an enanthem in the mouth are listed in Table 4.5; in all these infections, virus is likely to be present in the saliva. **Herpes simplex** virus is often shed intermittently in the absence of symptoms and presents a risk of herpetic whitlow to the operator. Cytomegalovirus is also shed in the absence of symptoms, but does not present a hazard. As far as viral infections are concerned, hepatitis B and HIV infections give the greatest cause for concern.

Hepatitis B (Chapter 20). The blood and oral secretions of HBeAg-positive but symptomless carriers may be infective. Recorded cases of dentists acquiring this infection from patients are rare, but the risk is significant and the appropriate precautions must be taken. It is advisable for all dental care staff having direct contact with patients to receive hepatitis B vaccine.

HIV (Chapter 22). Patients with full-blown AIDS often have fungal infections of the mouth and the characteristic purplish lesions of Kaposi's sarcoma. Viral infections include chronic herpetic stomatitis, hairy leucoplakia of the tongue (possibly due to EBV infection), and oral condylomata caused by papillomaviruses. Any of these appearances should arouse strong suspicion of AIDS, if it has not already been diagnosed. The transmissibility of HIV is substantially less than that of HBV, but the precautions to be taken are the same.

The golden rule for dentists and oral hygienists is

Always **wear rubber gloves when operating in the mouth**.

3.2   *Infection of patients by dentists*

There have been several transmissions of hepatitis B to patients from HBeAg-positive dentists, and a probable transmission of AIDS. Dentists suffering from either of these infections are not necessarily debarred from contact with patients, provided adequate precautions are taken, but obviously, any dental worker suffering from either infection must follow the advice given by an appropriate expert.

3.3   *Disinfection and sterilization of instruments, prostheses, etc.*

The main considerations here are prevention of cross-infection from patient to patient and from patient to dental laboratory staff. Prostheses and some delicate dental instruments cannot be treated by heat, the most effective method, and recourse must be had to glutaraldehyde (see Section 4). Disposable items are, of course, preferred wherever possible.

4   DISINFECTION AND STERILIZATION

Disinfection is the microbiological decontamination of infective material; preliminary cleansing of re-usable items, such as instruments, is an essential part of the process. **Sterilization** is the destruction of all infective microbes on a clean article so that it may safely be used for clinical purposes.

Except for the 'unconventional agents' (see end of section) viruses are inactivated by the same procedures as those used for other microbes. Heat is by far the most effective method; typical treatments are autoclaving at 121°C for 20 min or hot air oven at 160°C for 60 min.

Disinfectant solutions

These are used only when heating is not possible, e.g. for treating contaminated surfaces, pipettes, delicate instruments, etc.

The phenolic disinfectants used to kill bacteria are less active against viruses, which are, however, readily inactivated by hypochlorite solutions and — rather less effectively — by glutaraldehyde.

Hypochlorite solutions are highly corrosive; gloves and aprons should be worn when making up or using them and they should not be employed on metal equipment. The following working concentrations should be used.

Decontamination of spills of blood or body fluids	10 000 parts per million (p.p.m.)
Decontamination of potentially (i.e. not visibly) contaminated surfaces	1000 p.p.m.

Glutaraldehyde solution. Objects to be treated should be first cleansed in detergent and hot water, since glutaraldehyde does not penetrate coagulated protein. They are then immersed in a **freshly prepared** (same day) 2 per cent solution for at least an hour but preferably overnight. Glutaraldehyde fumes are toxic so that good ventilation is essential when preparing or using it and all containers must be closed. Rubber gloves must be used when handling it.

'Unconventional agents' (Chapter 23)

The agent of Creutzfeldt–Jakob disease (CJD), is, like others in this group, unusually resistant to all forms of disinfection. In the UK, the Department of Health and Social Security have issued guidelines for dealing with potentially contaminated instruments and tissues. They include a recommendation that such material should be autoclaved with

one cycle at 134°C for 18 min holding time,
  or
six cycles each at 134°C for 3 min holding time.

VIRAL INFECTIONS NOTIFIABLE IN THE UK

Viral infections

Acute poliomyelitis
Lassa fever
Marburg disease
Measles
Mumps
Rabies
Rubella
Viral haemorrhagic fevers
Yellow fever

Possibly caused by a virus

Acute encephalitis
Acute meningitis
Food poisoning
Infective jaundice

SUGGESTIONS FOR FURTHER READING

SUGGESTIONS FOR FURTHER READING

This book supplies all the information needed both for passing the final undergraduate examination in virology and for general clinical practice, but we recognize that, in view of the increasing importance of the subject, some of our readers will want to explore various aspects in greater depth. The following list includes both large reference works obtainable in most good medical libraries and some shorter texts that students may wish to buy for themselves.

General virology

Topley and Wilson's Principles of bacteriology, virology and immunity, 8th edn Vol. 4, Virology, (ed.) M.T. Parker and L.H. Collier (1990) Edward Arnold, London.
A comprehensive and up-to-date account of general virology, individual viruses and their diseases.

Zuckerman, A.J., Banatvala, J.E., and Pattison, J.R. (1987).*Principles and practice of clinical virology*. John Wiley & Sons, Chichester.
Covers the individual viruses and their diseases but not their general properties. Good on the clinical side.

Evans, A.S. 3rd edn (ed.) (1989). *Viral infections of humans: epidemiology and control*, Plenum, New York.

This is a first-rate book on the epidemiology of viral infections; it also contains much general information about the viruses themselves.

Fields, B.N., Knipe, D.M. (ed.) (1990). *Field's Virology* 2nd edn. Raven, New York.
A very large reference work (2560 pages) in two volumes.

Singleton, P. and Salisbury D. (1987). *Dictionary of microbiology and molecular biology* 2nd edn. John Wiley & Sons, Chichester.
Concise and authoritative monographs on the topics covered by the title. Expensive, but a mine of information.

Congenital infections

Remington, S.J. and Klein, J.O. (ed.) (1983). *Infectious diseases of the fetus and newborn infant* 2nd edn W.B. Saunders, Philadelphia
A comprehensive account, in one of the few textbooks devoted to this subject.

Laboratory aspects

Mahy, B.J.W. (ed.) (1985) *Virology: a practical approach*, IRL, Oxford
A detailed bench manual of procedures such as the propagation, purification, and assay of viruses. More useful for the research worker than in the diagnostic laboratory.

Doane, F.W. and Anderson, N. (1987). *Electron microscopy in diagnostic virology: a practical atlas and guide*. Cambridge University Press, Cambridge.
This book provides detailed information about the techniques of electron microscopy, e.g. negative staining, ultra-thin sectioning etc., but not about the electron microscope itself.

Antiviral therapy

Galasso, G.J., Whitley, R.J., and Merigan T.C. (ed.) (1990). *Antiviral agents and viral diseases of man* 3rd edn Raven Press, New York.
A comprehensive and reasonably up-to-date account. Further editions are to be expected as the subject expands.

Immunology

Roitt, I., Brostoff, J., and Male, D. (ed.) (1989). *Immunology* 2nd edn. Churchill-Livingstone, London
The importance of immunology to our understanding of the pathogenesis and control of viral infections cannot be overstressed. This book is an excellent pre-

sentation of basic immunology, well illustrated with line drawings and some photographs.

Safety precautions in dentistry

Macfarlane, T.W. and Samaranayake, L.P. (1989). Clinical oral microbiology. Wright, London.

British Dental Association (1987). *Guide to blood-borne viruses and the control of cross-infection in dentistry.* British Dental Association, London.

Index